Video Editing and Post Production

James R. Caruso
Mavis E. Arthur

Prentice Hall
Englewood Cliffs, New Jersey 07632

Library of Congress Cataloging-in-Publication Data
Caruso, James R.
    Video editing and post production / James R. Caruso, Mavis E.
Arthur.
        p.   cm.
    Includes bibliographical references and index.
    ISBN 0-13-946575-8 (paper). — ISBN 0-13-946583-9 (case)
    1. Video tapes—Editing.   I. Arthur, Mavis E.,
II. Title.
TR899.C37   1992
778.59'92—dc20   091-22402
                                                      CIP

Editorial/production supervision: *Harriet Tellem*
Cover design: *Bruce Kenselaar*
Prepress buyer: *Mary Elizabeth McCartney*
Manufacturing buyer: *Susan Brunke*
Acquisitions editor: *George Kuredjian*
Editorial assistant: *Barbara Alfieri*
Illustrations: *Mary LaPorte*
Original photographs: *James R. Caruso and Mavis E. Arthur*

 © 1992 by Prentice-Hall, Inc.
A Simon & Schuster Company
Englewood Cliffs, New Jersey 07632

Photographs produced by the authors were printed from videotape on a Hitachi color video printer.

The publisher offers discounts on this book when ordered
in bulk quantities. For more information, write:
    Special Sales/Professional Marketing
    Prentice-Hall, Inc.
    Professional & Technical Reference Division
    Englewood Cliffs, New Jersey 07632

Printed in the United States of America
10 9 8 7 6 5 4 3 2 1

ISBN 0-13-946583-9 {C}

ISBN 0-13-946575-8 {P}

Prentice-Hall International (UK) Limited, *London*
Prentice-Hall of Australia Pty. Limited, *Sydney*
Prentice-Hall Canada Inc., *Toronto*
Prentice-Hall Hispanoamericana, S.A., *Mexico*
Prentice-Hall of India Private Limited, *New Delhi*
Prentice-Hall of Japan, Inc., *Tokyo*
Simon & Schuster Asia Pte. Ltd., *Singapore*
Editora Prentice-Hall do Brasil, Ltda., *Rio de Janeiro*

Contents

7   Advanced Editing Systems   **212**

8   Recording and Editing Audio   251

9   The Edit Suite   270

Preface

The hardest and yet most rewarding part of the video process is post production and editing. It is hard because it is the last chance to make your video as great as you thought it would be. It is a time of realizations, of accepting the show for what it is—good or just adequate. It is rewarding because it brings you to the conclusion of the process and that alone may be reward enough. If the video goes beyond your expectations, that is a bonus.

This book is about all aspects of the post production and editing process, from the basic planning to the equipment to the edit itself. It deals with the answers to questions such as these:

What is the language used in post production?

How does TV work?

How is TV edited?

What is an EDL?

How do you hook up your equipment for maximum effectiveness?

How do you select videotape?

How do you set up a basic editing system?

What is time code?

What is the difference between assemble and insert editing?

How do you add audio?

How do you make transitions from one shot to the next?

How can you be selective with your footage?

And what do you do with 10 hours of prerecorded tape?

While you might not know everything about editing when you complete this book, you will be able to make well-informed choices concerning the editing process. At the very least, you will have a basic knowledge of effective post production and editing.

While preparing this book, we called on friends and manufacturers to make it as comprehensive and up to date as possible. We would particularly like to thank the following people and manufacturers for their help: Bob Cohen, President, FutureVideo Products; Sam Sunshine, Product Specialist, and Mikey Sekiya, Planning Specialist, JVC Company of America; Victoria Chaffee, District Sales Manager, Panasonic AV Systems Group; Sue Robbins, Vice-President, Jean Doynow Associates; Richo Company Ltd.; Sylvia Waters, Manager, Marketing Administration, Sima Products Corporation; Alex Shumasu, Product Planning, and Tony Sudo, Product Manager, Sony Corporation of America; Dennis Vance, President, Vertex Video Systems; Stan Pinkwas, Managing Editor, *Video Magazine;* and Darrel Gore, Product Manager, Videonics, Incorporated.

All photographs in the book (except manufacturer's products) were printed directly from the video signal using the Hitachi video printer. This fine piece of equipment will also reproduce photos in full color, just as you see them on the screen. Our special thanks to these people at Hitachi Home Electronics: Eric Kamayatsu, Director of Sales, Multi Media Systems; Walter Lockhart, National Advertising Coordinator; and last but not least Steve Shattuck, Account Executive, William Campeau Public Relations.

Introduction

Editing is by far the most rewarding and, at the same time, the most painful part of the creation of a television program. It is fun and frustrating at the same time. It is a time of final decisions, of inevitable compromises, of painful realizations and, ultimately, a time of overwhelming rewards when the deed is accomplished and the final edit is made.

Inevitably, the "could-ofs," "should-ofs," "if I'd only" set in and you begin to pick apart your work of art, exposing all its flaws until it becomes less than what it was, at least to you. You may think of reediting, even reshooting, or just scrapping the whole thing. Somewhere along the way, you lost what the program was supposed to be, allowing yourself to be caught up in its minute detail. You know so much about *how* it was made that you have forgotten *why* it was made. Worst of all, you have lost your objectivity and any flaw, however minute, causes you to cringe. When this happens, the only thing to do is "walk away." Let the show rest and give yourself time to forget the details so that you can appreciate the whole.

Editing is not easy. It is your last contact with something you have created, and sometimes it is hard to let go—to say "it's good enough, it's great, it's the best I've ever done." Or, at least, "it's as good as it can be." There's always the temptation to do just one more thing, to make one more edit, to go shoot one more pickup, to incorporate just one more sound effect.

Editing is not easy but it *is* worth it. If you understand the pitfalls, if you understand that there is a point where you can do no more, if you know the abilities of your equipment, then you can make your video the best it can be. And while the temptation to make just one more edit will probably always be there,

there will be a point where you can walk away from the process and enjoy the show, just like your audience.

This book is meant to help you reach this point. It is designed to walk you through the frustrations and rewards of editing—to help you learn the mechanics and techniques of television's most creative aspect, editing.

1

Post Language

The mechanics and the creative, these are the two most important elements of post production. A total understanding of both is necessary to transform raw footage into an effective program. Both will be dealt with in the pages of this book, but first it is important to learn the language of post.

Television production has devised its own way of saying things, a shorthand if you will, to easily describe the various steps, techniques, and technical aspects of the process. The following glossary will help demystify the post-production process and prepare you for dealing with the mechanics and techniques of creating a television program as presented within the pages of this book. Read it now as an overview and refer back to this listing if you have questions as you read.

GLOSSARY

A (audio) edit An edit that records audio only.

A roll The first-generation recorded camera original that contains all the video and audio recorded.

Action What happens in the shot, both video and audio. In other words, the pictures and the sound that tell the story of a particular shot.

Aspect ratio The relationship between the height and width of the television screen. In TV the standard aspect ratio is 3:4, meaning that if the screen is 3 inches high, it will be 4 inches wide or multiples thereof. See Figure 1–1.

Assemble edit To put the shots together one right after the other, in sequential

1

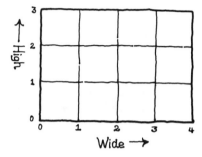

Figure 1-1 TV aspect ratio is 3:4. (James R. Caruso/Mavis E. Arthur, A BEGINNER'S GUIDE TO PRODUCING TV© 1990, p. 3. Reprinted by permission of Prentice Hall, Englewood Cliffs, New Jersey.)

order from the beginning to the end of the video. Every time an assemble edit is performed it replaces all previously recorded material with new video, audio, and control track. See Figure 1-2.

Audio The sound portion of the video recording, including voice, music, and sound effects. All videotape has the ability to record one, two, three, or four separate audio tracks, depending on the capabilities of the VCR.

Audio sweetening The process of mixing the audio recorded on the edited master which may include any or all of the following: additional or dubbed-in dialog, voice over announcer, sound effects, music, and sometimes even a laugh track.

Auto assemble Allowing a computerized edit system to put together a program

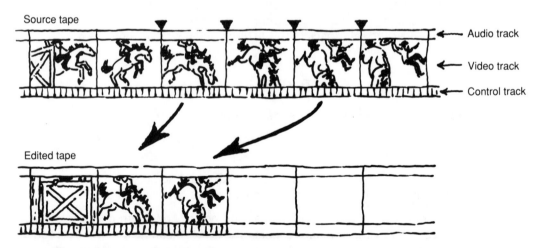

Figure 1-2 Assemble editing records the shots sequentially, replacing all previously recorded material including video, audio, and control track.

in an on-line edit session based on an edit decision list (EDL) created during an off-line edit session.

A/V edit   An edit that records both audio *and* video.

B roll   The secondary shots and/or audio to be used as cover or cutaways for the video. It may also be shots duplicated from the A roll material to create source material for a dissolve or other effect.

Back time   The method of calculating the edit-in point (where the edit will begin) by subtracting the length of the shot from the edit-out (where the edit will end) point. This is done when it is critical that the video material being edited begin and end at a specific point, such as in an insert edit.

Backspacing   Rewinding the tape a specific amount from the edit-in point to give the VTR or VCR enough time to get up to speed before recording begins at the edit in. A machine backspaces so it can preroll. See also *preroll.*

Bias   Makes sure that the videotape is magnetized over the range of the recorder's head magnetic characteristics.

Blacked edit master (BEM)   Videotape that has black, control track, and sometimes time code already recorded on it. In computerized editing, a blacked tape gives the computer a reference point.

Breakup   A disturbance in the video or audio signal caused by the loss of sync, videotape damage, or electrical interference. Also known as *glitch.* See Figure 1–3.

Bulk   Erasing a tape completely, removing any and all video and audio on the tape. Tapes are bulked by running them over a magnetic field. See *degauss.*

Bumper   Used at the beginning or the end of a video segment; a bumper gives the viewer additional information, for example, about another program, about what is going to happen next, or simply the name of the program itself. See Figure 1–4.

Bumping up or down   Transferring video from one format to a larger one (8mm to $\frac{1}{2}$-inch, or $\frac{1}{2}$ inch to $\frac{3}{4}$ inch, or $\frac{3}{4}$ inch to 1 inch) *or* to a smaller one (1 inch to $\frac{3}{4}$ inch, or $\frac{3}{4}$ inch to $\frac{1}{2}$ inch, or $\frac{1}{2}$ inch to 8mm).

Burn.   To erase or record over existing material. For example, you may elect to burn an edit you made, erasing it completely, or burn a shot you recorded while shooting.

Burn in   A spot or streak on the video caused when the camera is pointed too long at a bright object, for example, the sun.

Camera angles   The angle of the camera in relationship to the shot. The camera angle can be high, shooting down, making the performer or action appear

Figure 1-3 Breakup or a glitch on the recording is caused by a disturbance in the video or audio signal, for example, loss of sync, videotape damage, or electrical interference.

Figure 1-4 Bumpers appear at the beginning, before, or after commercial breaks or at the end of a program as an ID or to provide additional information.

small and insignificant. See Figure 1–5. The camera angle can be low, shooting up, making the performer or action seem huge and overbearing. See Figure 1–6. Or it can be straight forward or head on from a normal perspective. See Figure 1–7. In addition a camera can be shooting from one of three perspectives: (1) *first person* point of view (POV), where the viewer is put directly into the shoes of one of the performers (see Figure 1–8); (2) *second person,* where the viewer stands alongside the performer, participating but not directly (see Figure 1–9); and (3) *third person,* where the viewer is removed from the action and sees it only as an observer (see Figure 1–10).

Camera framing The framing of a shot is designated based on the size of the main subject matter in relationship to the picture itself, as follows:

EWS (extreme wide shot) The widest shot possible of the location, that is,

Figure 1-5 High-angle shots make the subject appear smaller and insignificant. (James R. Caruso/Mavis E. Arthur, A BEGINNER'S GUIDE TO PRODUCING TV© 1990, p. 60. Reprinted by permission of Prentice Hall, Englewood Cliffs, New Jersey.)

Figure 1-6 Low-angle shots make the subject seem huge and overwhelming. (James R. Caruso/Mavis E. Arthur, A BEGINNER'S GUIDE TO PRODUCING TV© 1990, p. 60. Reprinted by permission of Prentice Hall, Englewood Cliffs, New Jersey.)

a panoramic long distance shot. On television, an EWS is used to establish setting and location since it is generally too wide to be read effectively on the small screen. See Figure 1-11.

WS (wide shot)   A variation on the EWS, this is used as an establishing shot or to forward the action, since it allows the viewer to recognize distinguishing details and features. See Figure 1-12.

MS (medium shot)   A half-shot of the object or person in the scene. See Figure 1-13.

CU (closeup)   A close shot of a person or object. See Figure 1-14.

ECU (extreme closeup)   An even closer shot. See Figure 1-15.

Camera moves   The camera can move from one location to another or its lens can move. The camera moves in five basic ways. See Figure 1-16.

Zoom   The camera lens can zoom in or out by changing its focal length.

Figure 1-7 Head-on (straightforward) shots give the viewer a normal perspective on the subject. (James R. Caruso/Mavis E. Arthur, A BEGINNER'S GUIDE TO PRODUCING TV© 1990, p. 60. Reprinted by permission of Prentice Hall, Englewood Cliffs, New Jersey.)

Zooms are also sometimes referred to as a pull (zoom out) and a push (zoom in). A zoom can be slow or fast. A pop zoom is an extremely fast zoom, popping in or out.

Truck The entire camera moves left or right parallel to the main subject in the shot. It can be moved in a variety of ways, including on wheels (on a dolly or mounted on an automobile) or strapped to the cameraperson, (on a Steadicam rig or on a crane).

Dolly The camera moves closer to or farther away from the main subject in the shot.

Pan The camera moves left or right on its pedestal in a sweeping motion. The camera itself does not change its location but rather it scans the scene either left to right or right to left.

Tilt The camera tilts up or down on its pedestal. The camera itself does not change its location but remains in its fixed position, tilting only its head.

Figure 1-8 Shooting from a first Person or POV (point of view) puts the viewer directly into the shoes of one of the performers.

Figure 1-9 A second-person camera angle puts the viewer alongside the performer, allowing the viewer to participate, but indirectly.

Figure 1-10 A third-person camera angle looks at the subject as an observer, removed from the action.

Figure 1-11 An extreme wide shot (EWS), the widest shot possible of the subject, is used to establish setting and location.

Figure 1-12 A wide shot (WS) can be used to establish location *or* to forward the action.

Figure 1-13 A medium shot (MS) isolates an object. A WS of a train yard, for example, cuts to a MS of one train.

Figure 1-14 A MS of the train cuts to a closeup (CU) of the wheel and gears of the train.

Figure 1-15 A CU of the wheel and gears cuts to an extreme closeup (ECU) of one wheel.

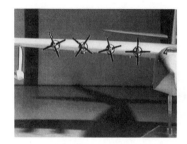

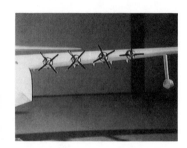

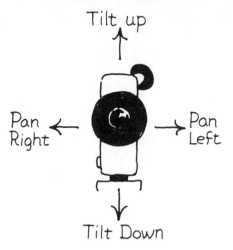

Tilt up

Pan Right ← → Pan Left

Tilt Down

Truck Right ← → Truck Left

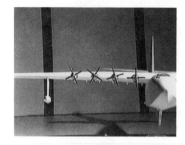

Figure 1-16 The camera moves in four basic ways: zoom, truck, pan, and tilt. Zoom, pan, and tilt are illustrated in the drawings. Pan is also visually illustrated in the shots of the airplane at the top of the page. Truck is illustrated in the series of photographs of the airplane at the bottom of the page. (James R. Caruso/Mavis E. Arthur, A BEGINNER'S GUIDE TO PRODUCING TV© 1990, p. 20. Reprinted by permission of Prentice Hall, Englewood Cliffs, New Jersey.)

Character generator   A computer specifically designed to electronically generate letters, numbers, and symbols to be included in a video production for titles, credits, or information within the production. Sometimes called a *CG*. See Figure 1–17.

Chroma key   A method of laying one video picture over another electronically by using any one of three primary colors (red, blue, or green) as a background for the picture that will be overlaid and then processing the signals from both pictures through a special-effects generator. The most common background color is blue, then green. Also known as a *key*.

Chrominance   The color information (saturation and hue) of the video signal. Also known as *chroma*.

Close   The final video and/or audio in the program, not including any credits.

Color balance   The adjustment of the primary colors using color bars as an electronic reference to obtain a desired balance among the colors, all being as close to standard as possible. Color balance can be achieved manually by looking at the colors, *or* it can be achieved electronically by white balancing the camera for existing lighting conditions (see *white balance*).

Color bars   Standard color test signal or pattern, which includes rows of bars

Figure 1–17   A character generator is a computer designed to electronically generate letters, numbers, symbols, and some prebuilt graphics. (Equipment courtesy of Videonics.)

colored, in this order: gray, yellow, cyan, green, magenta, red, and blue. See Figure 1–18.

Colorize Adding a color to a picture or a graphic, for example, making the title blue.

Component video A video system that records the luminance (black and white) and sync information separately from the color information.

Composite video A video signal that combines chrominance, luminance, and sync information.

Continuity Continuing the sequence or story logically and accurately and keeping the details consistent from one shot to another, including screen direction, screen position, and matching action and wardrobe and surroundings. For example, if the performer wears a hat in one shot of a scene and is walking toward the saloon, when the camera cuts to a different angle of the same scene, the performer should still be wearing the hat and walking in the same direction. If the story line follows the solving of a murder, the solution should proceed in a logical manner, not jump about or skip critical elements. Continuity of lighting and audio levels are also important to a well-made video.

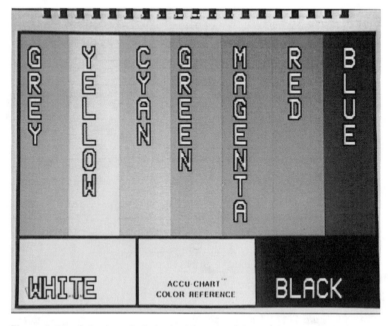

Figure 1-18 Color bars include, in this order left to right, gray, yellow, cyan, green, magenta, red, and blue.

Control track Electronic pulses recorded on the edge of the videotape that control the longitudinal velocity of the videotape, aligning and synchronizing as well as counting the pulses recorded on the videotape during playback. The information can be read out with some types of VCR counters in hours, minutes, seconds, and frames. Control track is to videotape what sprocket holes are to film. See Figure 1–19.

Control track editing An editing system that uses the frame pulse on the control track to count frames to set accurate edit-in and edit-out points. Control track can read out in time code-like numbers, even though it is not recorded time code.

Countdown Keeping track of the time remaining before action begins or a video ends; for example, a 30-second commercial counts down to 30 seconds since it has only that allotted time to tell its story. This term is also used for the countdown to first video; for example, 10 appears on screen, then 9, then 8, and so on down to 2; then there is 1 second of black and then the video begins.

Counter numbers The counter may be a mechanical device that simply measures the length of the tape as it is run through a VCR. Some counters use the control track to determine approximate real time. No information is recorded on the videotape; therefore, a camera counter must be zeroed at the beginning of a videotape, and any VCR that that tape is played back on must be zeroed at the beginning of the tape in order for the numbers to mean anything. At best, counter numbers will get you close to a desired shot, but not exactly on it, since the mechanics of the counter function will vary from machine to machine, and the zero point may vary as well. See Figure 1–20.

Cover or cutaway A shot specifically inserted to either explain a part of the action or to provide a transition to the next scene; for example, someone

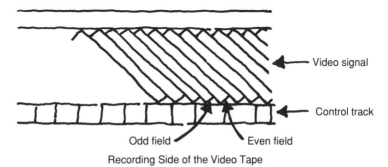

Recording Side of the Video Tape

Figure 1–19 Control track is a series of electronic pulses, one for every frame, recorded on the edge of the videotape.

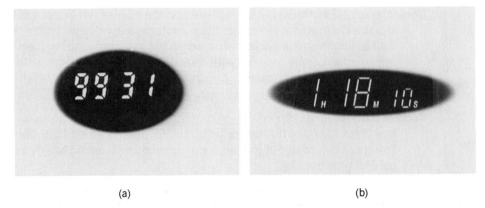

(a)   (b)

Figure 1-20 (a) Counter numbers are purely mechanical, counting in sequential numbers. (b) Control track numbers count frame pulses, reading out like time code in hours, minutes, seconds, and, in some cases, frames.

moves from one building to another, and the director may call for a cutaway of cars driving on the freeway. This shot may also be used to indicate the passage of time. See Figure 1–21.

Credits Names of those individuals or companies that worked on or assisted in the video, for example, the director, writer, and performers, and the locations used. Some credits may appear at the beginning of the program, but most will appear at its conclusion. See Figure 1–22.

Cue The exact location on the tape where a video or audio segment begins; an edit-in point.

Cueing up The interval of time when the computerized editing machine is looking for a desired location, usually designated in time code numbers. When it finds the location, it is cued up and parked (stopped or paused) until the editor or the computer tells it what to do next.

Cuts only editing A function of the equipment's capabilities; an edit that cuts from one shot to the next. A cuts only edit system is two VCRs: one to playback, the other to record. No other transitions such as dissolves and wipes are possible.

Degauss To demagnetize the tape, erasing all recorded material. See also *bulk*.

Digital video effects (DVE) A memory device that stores a frame of video information in a digital circuit, converting it from analog to digital form so it can be manipulated to create a variety of different special effects, for example, rolling, tumbling, sliding, or breaking up. See Figure 1–23.

Dolly See *camera moves*.

(a)

(b)

(c)

Figure 1-21 Insert a cutaway between two shots to explain or further the action or to act as a transition.

Drop Frame Time Code A time code format that drops or skips two frames every minute except on the tenth minute to make it correspond to real time. A program edited with drop frame time code will be 18 frames shorter than one edited with nondrop time code, and yet be the same length in real time. See also *non drop frame time code*.

Dropout A loss of recording surface or oxide on a piece of videotape causing holes in the picture; a picture with dropout has black or white spots on it. See Figure 1-24.

Dub Rerecording video and/or audio onto another tape, thus making a duplicate copy.

Dub master A dub (duplicate) of the edit master and generally used to make

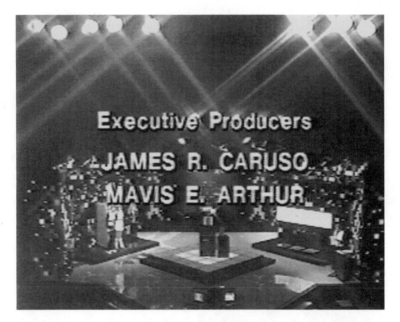

Figure 1-22 Credits recognize those individuals who helped produce the program.

multiple copies for distribution. Thus the edit master is protected from any damage that could occur from repeated play. A dub master is usually a third-generation videotape, the edit master is second generation, and the original material is first generation. Also known as a *protection master.*

Edit Adding graphics, sound effects, music, and whatever else to the original video and audio recordings and putting it all into a logical order to tell the story as you have conceived it. Simply, it is transforming hours of real-time video into a concise, effective video that tells the story. Also known as *post.*

Edit in The exact point on the edit master where the new video and/or audio will begin recording.

Edit decision list (EDL) A list of edits made in an off-line edit session used to autoassemble a program during an on-line edit session. See Figure 1-25. An EDL may also be a paper edit, a list of all the shots that you have decided to use in the program by time code or counter number, with transitions between them indicated, as well as the insertion of any graphics, special effects, sounds, or music. See Figure 1-26. In other words, an EDL is the complete program reduced to paper.

Edit master The edited program.

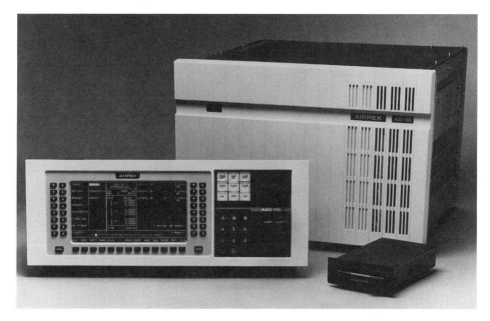

Figure 1-23 Digital video effects are created by converting video information from analog to digital so that the signal can be manipulated to create pictures that tumble, roll, slide, and so on. Pictured is Ampex's ADO 100. (Photo courtesy of Ampex Corporation.)

Edit out The exact point where the material being recorded on the edit master will end.

EFX Short for effect; used to designate a visual effect (see also *SFX*).

Enhancing Electronically improving the look by increasing or decreasing the color or adjusting the color balance and sharpness of the video image.

Establishing shot Generally, an extreme wide shot (EWS); it establishes setting and location, for example, the exterior of a hospital to give added meaning to a surgery scene. See also *Camera framing, EWS.*

Field Half of one television frame. A frame of video (NTSC $= \frac{1}{30}$ of 1 second) is created in two separate scans. The first scan or field creates the odd-numbered lines in the frame of video, while the second scan, field 2, creates the even-numbered lines. See Figure 1-27.

First generation Raw footage and master material; this is the tape on which the video camera recorded the video and/or audio originally. Also known as *camera original.* See *generation.*

First video The very first visual seen on the program.

Figure 1-24 When tape loses a part of its magnetic coating, dropout appears. In this illustration, dropout is shown over video black.

Flash frame One field or one frame of unwanted video.

Flying erase head Erase head mounted on the same drum as the video heads. It erases video and/or audio immediately prior to an insert edit. If it is an assemble edit, it will erase the control track in addition to the video and audio information.

Frame One complete video picture. It takes 30 frames to make 1 second of video by NTSC standard and 25 frames per second by PAL and SECAM standard.

Frame accurate To be able to access a single frame of video, which represents $\frac{1}{30}$th of a second (NTSC).

Freeze One frame of video frozen in place just like a photograph. Also called a *still*.

Generation A designation used for tape to indicate how many times the original has been dubbed. The original material is always first generation. The edit master recorded from this first-generation material is second generation. A dub made from the edit master is third generation, and so on.

Genlock An electronic device that listens to one video signal and then manufactures additional signals in step (in sync) with the original. Used for switch-

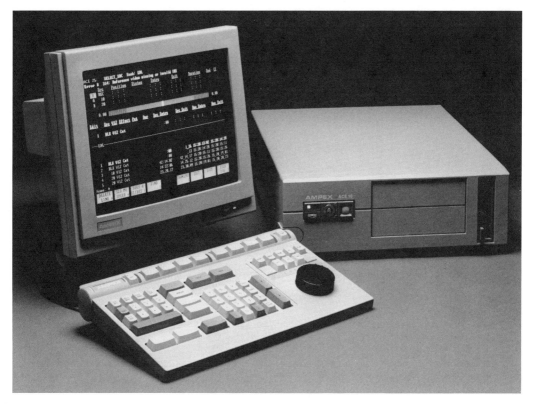

Figure 1-25 An edit decision list (a computerized version is shown here on the monitor) can be created on a computer during an off-line session and used later in the on-line edit session. (Photo courtesy of Ampex Corporation.)

ing various video input devices cleanly, such as from camera to camera or computer to VCR. See Figure 1-28.

Glitch   See *breakup.*

Glitz   This is that intangible, abstract something that gives the video sparkle, making it come alive and be memorable for the viewer. Also called *pizazz.*

Graphics   Any handmade or computer-generated artwork or letters included in the video.

Helical scan   A type of recording on a VTR that wraps the videotape partially or completely around the recording head, causing the video signal to be recorded across the tape in a slanted pattern. See Figure 1-29.

Hit   Interference or breakup on one frame or as little as one field; a minute flaw. Also called *flash.*

EDIT DECISION LIST

| Scene | Tape No. | Shot No. | Edit type | Transition | Time code/counter numbers | | Duration | Description |
| | | | | | In | Out | | |
					(tape no. is designated hour)			
	6		AI	FADE UP	00:21:23:25	00:21:25:22	01:27	MUSIC
23	7	23-3	V	FADE UP	02:10:35:10	02:10:36:20	:30	CU2SHOT
24	7	24-7	V	C	02:18:22:11	02:18:23:26	:45	WS

Figure 1-26 An edit decision list (EDL) could be a paper edit, a list of all shots prepared by hand for use in the edit session.

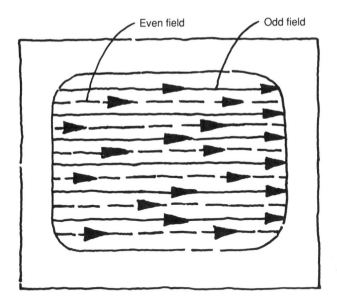

Even field   Odd field

Figure 1-27 One picture or frame of video is created when two fields interlace.

Figure 1-28 A genlock is an electronic device that syncs up various sources. (Photo courtesy of Digital Creations, Inc.)

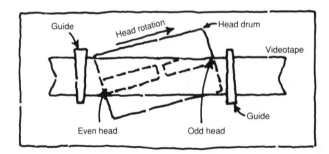

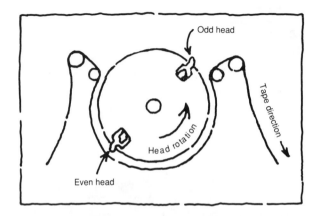

Figure 1-29 Helical scan records the video signal in a slanted pattern.

ID Several seconds of video and/or audio that identify for the viewer what they are watching or where they are watching it; for example, on commercial television a program ID may be inserted between commercials or the station may put up its call letters and/or channel number.

Insert edit Placing video and/or audio between two shots already recorded, thus erasing any video and/or audio already recorded there. This type of edit does not record new control track, but leaves previously recorded control track in place. See Figure 1–30.

ISO A camera or VCR devoted to only one part of the action or story, for example, a camera shooting only one horse in a horse race.

Jogging Moving the video forward or backward one frame at a time.

Key An electronic method of inserting graphics over a scene, thus superimposing them or putting one image over another. Also called a *luminance key*.

Keystone A slanting effect by *not* shooting a graphic with the camera on the

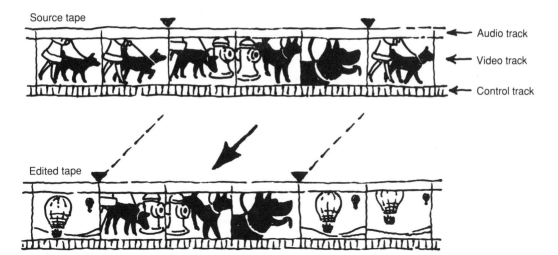

Figure 1-30 Insert editing records audio and/or video between existing video and/or audio, but leaves the recorded control track in place.

exact plane with the artwork, making the art appear closer at the closest point to the camera and at a distance at the farthest. It can be a mistake or an effect. See Figure 1-31.

List management Allows an editor to shift or manipulate edits stored in the computer's memory. See also *edit decision list*.

Longitudinal time code (LTC) Time code that records on one of the audio tracks of the videotape. See also *vertical interval time code.*

Luminance The brightness of the black and white portion of the video signal indicated in amplitude (strength) as measured on a gray scale. In other words, the brightness levels of the picture from reference black to peak white.

Match frame Performing an edit in sync with the same video as the previous shot or edit to lengthen a shot or to begin a transition such as a dissolve. Simply put, a match frame is finding the exact same frame already recorded

Figure 1-31 Keystoning occurs when a graphic is shot on a slant, not straight on, making it appear closer at one end and farther away at the other end.

and beginning the recording at exactly that same frame so that there is no break in the video.

Matte   A key effect that replaces light parts of the video with a color; for example, by replacing them with blue, the picture becomes blue intensive. Matting is usually associated with adding color to black and white graphics, that is, titles and credits.

Montage   Combining a variety of different shots by editing them together, with each appearing only a short period of time (seconds or fractions of seconds) to create a strong although sometimes abstract vision. Montage sequences are the basis for many music videos by combining shots of a performance with abstract images and/or action related to the words of the song.

MOS   Recording video without audio, "mit-out-sound."

Noise   Any element that interferes with the clarity of the video picture, such as insufficient light at the time of the original recording, electrical interference, or dubbing down too many generations. See Figure 1–32.

Non drop frame time code   A time code format that counts a full 30 frames a second, making it longer than real time by 108 frames every hour. This time discrepancy is corrected by calculating the difference and making the ad-

Figure 1-32   Insufficient light at the time of the original recording can create noise in the picture.

justments to the edit master or by using drop frame time code which compensates automatically. See drop frame time code.

NTSC (National Television Standards Committee) The color video system in which the electronic beam scans the picture tube at a rate of 525 lines of picture information 30 times every second. In other words, the picture is created one line at a time, with 525 lines making one complete picture every $\frac{1}{30}$ second. The NTSC system is standard for television in the United States as well as Canada, Chile, and Japan (see also PAL and SECAM).

Off-line editing A preedit done to establish editing points and create an EDL. It usually includes only rough transitions, if any, and may or may not include all the elements of the program, for example, graphics and special effects. This edit is designed to save time and money when the final edit is done by making it possible to do an autoassemble edit using the EDL.

On-line editing The final edit with all the "bells and whistles" available to complete the program as you had planned, that is, effects, music, graphics, and original taped material. It may use an EDL generated in an off-line edit or it may start with only a paper EDL.

Open The beginning action of the program; it may include the title sequence if the action begins over titles and opening credits.

Open-ended edit Making an edit without designating an out point so that the recording continues until the editor manually stops it.

Outtake A portion of the original video recording (raw footage) that is interesting, sometimes amusing, but not appropriate for the video itself. Since it is "too good" to discard entirely, you may create an outtake reel, which includes footage from one video only or from various videos, *or* it can be used under closing credits.

Pacing The apparent speed at which the story or sequence is told to the viewer. Also called *timing*.

PAL (Phase Alternating Line) A video color system in which the electronic beam scans the picture tube at a rate of 625 lines of picture information 25 times every second. A variation of the NTSC system, PAL was designed to eliminate some NTSC problems, specifically a shift in chroma phase (hue). The PAL system is standard in more countries than NTSC or SECAM, including Algeria, Argentina, Australia, Austria, Bangladesh, Belgium, Brazil, China, Denmark, Finland, Great Britain, India, Indonesia, Ireland, Italy, Malaysia, Norway, New Zealand, the Netherlands, Portugal, Spain, Sweden, Switzerland, Thailand, Turkey, and Yugoslavia (see also *NTSC* and *SECAM*).

Parked A tape that is cued up and stopped or paused at a desired location. See *cueing up.*

Playback machine   The machine that plays the videotape in an edit system. Also called a *source machine*.

Pop zoom   See *camera moves, zoom*.

Post   See *edit*.

Post production   All activity that follows the completion of taping up to the finished program, including but not limited to the selection of shots and deciding on transitions between those shots; the selection of music, audio, and sound effects; the selection of graphics and other visual elements; and, finally, the execution of these decisions in an edit session.

Posterization   Changing the brightness levels of a picture to give it a flat, painted, or etched look. See Figure 1–33.

POV   See *camera angles, first person*.

Preroll   The specified period of backspacing it takes the edit and source machines to obtain the proper head and tape velocity and sync to play back and record properly at the edit-in points. Preroll time is usually 10 to 12 seconds.

Preview   Looking at an edit before actually making it to be sure it is what you

Figure 1-33   A picture can be given a flat, painted, or etched look with posterization.

want to do. Also used to designate the monitor on which the source material can be seen while an edit is taking place.

Production cycle   All aspects of producing a video, from preproduction, which includes planning, scripting, casting, and scheduling, to the actual production, which includes taping, to post production, which includes planning and executing the edit.

Program   Designates the monitor on which the recorded edit can be viewed as it happens. The preview monitor is usually side by side with the program monitor.

Pull   See *camera moves, zoom.*

Punch and crunch editing   Basic electronic editing where the editor pushes the record button when the shot is visually seen. The editor literally punches the button and crunches the program together. The method is close, but not accurate, and a far cry from computerized time code editing.

Push   See *camera moves, zoom.*

Real time   Actual time; that is, letting a shot last as long as it takes for the action to really happen. For example, if driving across town takes 20 minutes, this sequence in real-time video lasts 20 minutes. Real-time video is usually boring to watch. Things just take too long. Viewers are trained to watch action condensed into TV time. See also *TV time.*

Record machine   The machine that is recording the videotape. In an edit session, it is creating the edited master.

Resolution   The amount of detail in the video image measured both on the vertical and horizontal axis. The higher the resolution, the better the picture quality is. In short, the more lines of picture information, the better the picture is.

RGB   Stands for red, green, and blue, the primary colors that make up color video.

Rough cut   Editing the master material together without all the elements, such as transitions and graphics. Think of it as a kind of rehearsal for the final edit. A rough cut may be used to help determine if the shots and scenes will cut together without continuity problems, or to discover the best order for the story, or to get approvals from executive producers.

Safe action area   The area of the television screen that is not in danger of being cut off by a misadjusted television set. Safe area is about 90% of the television screen measured from the center to the outer edges. If the action is *out of safe,* it is outside this area. See Figure 1–34.

Safe graphics area   Like the safe action area, this is the part of the television

Figure 1-34   Approximately 90% of the TV screen is within the safe action area. Any action outside this area *may* not be seen by the viewer.

screen that will always "go home" to a viewer's set. The safe graphics area is about 80% of the screen measured from the center. Any graphics outside this area is *out of safe* and might not be totally readable on a home TV. See Figure 1–35.

Saturation   The amount of color in the TV picture.

SECAM (Séquential Couleur á Mémoire)   A video color system in which the electron beam scans the picture tube at the rate of 625 lines of picture information 25 times every second. It was developed by the French and is used in France, the Soviet Union, and several other countries (see also *PAL* and *NTSC*).

SEG (special-effects generator)   A device designed to interface with VTRs and VCRs to create special effects for a program. These effects can include various wipe patterns, posterizing, matting, and keys. Some also have the ability to mix various audio sources. See Figure 1–36.

SFX   Abbreviation for sound effect; used as a designation that a sound effect is desired (see also *EFX*).

Show bible   All preproduction, production, and post-production plans and activities committed to paper and gathered together in one place, such as a

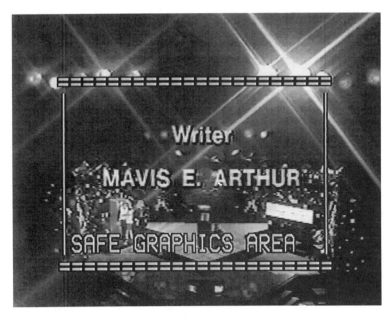

Figure 1-35 About 80% of the TV screen is within the safe graphics area. Any graphics out of safe *may* not be seen by the viewer.

three-ring binder, for ease in referring to them before, during, and after the shoot. Depending on which planning steps you choose to use, the bible may include a master checklist, schedule, show outline, budget worksheet, equipment list, crew list, scenario and format, script, storyboards, cast list,

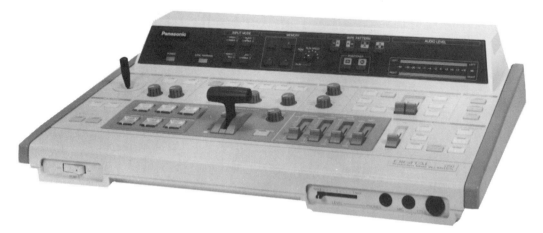

Figure 1-36 A special-effects generator (SEG) is designed to interface with VTRs or VCRs to create special effects for a program. Pictured is Panansonic's Digital Production Mixer MX 12. (Photo courtesy of Panasonic Audio/Video Systems Group.)

location list and sketches, prop list, shot list, shot log, edit list, and credit list, all tabbed for ease in locating. See Figure 1–37.

Shuttling   Going fast forward or in reverse to reach a desired position on the tape.

Slate   A board recorded at the beginning of every shot to help in relocating shots during the edit; it includes all pertinent information about a shot, for example, scene number, take number, the date, an indication of whether sound is being recorded, and the director and/or camera person. Slates are commonly used to locate shots in film and sometimes in video in conjunction with time code or counter numbers. See Figure 1–38. A slate is also used at the beginning of a program, to identify the program, the producer, the taping date and such. An example of this type of slate can be seen in Chapter Four, Figure 4–23.

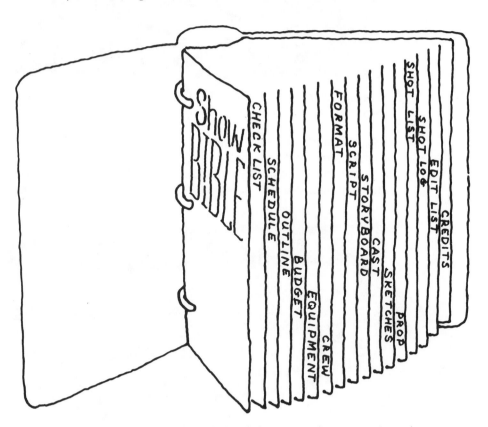

Figure 1-37   The show bible includes all the paperwork necessary to produce the show and can be kept in a three-ring binder. (James R. Caruso/Mavis E. Arthur, A BEGINNER'S GUIDE TO PRODUCING TV© 1990, p. 206. Reprinted by permission of Prentice Hall, Englewood Cliffs, New Jersey.)

Figure 1-38 A slate may be recorded on the tape prior to a scene to help relocate shots, supplementing the shot log. (James R. Caruso/Mavis E. Arthur, A BEGINNER'S GUIDE TO PRODUCING TV© 1990, p. 13. Reprinted by permission of Prentice Hall, Englewood Cliffs, New Jersey.)

SOT   Sound on tape, meaning that sound is recorded with the video.

Sound bite   Used liberally on newscasts; this is a brief commentary or an all-encomposing statement.

Speed   A verbal cue that the tape has been in play long enough for the heads to make contact and establish themselves so that there will be no glitch in the record or playback. Also called *up to speed.*

Split   A video effect that splits the screen into more than one picture. Two-way splits are common, for example, shot of a person talking on the telephone in the left part of the screen with a shot of the person on the other end of the line in the other half of the screen. A split may be used to give the viewer more information, as with the telephone conversation, or it may be used for effect if the split becomes so big. For example, in a 32-way split, the viewer can no longer see what is in individual pictures but, rather, the pictures become a pattern in themselves. See Figure 1–39 for an example of a split.

Split edit   Giving the audio and the video different in or out points for the edit; for example, the audio ends before the video does, or vice versa.

Steadicam⊕   The trade name of a camera rig that allows the cameraperson to mount the camera directly onto his or her body and move freely without causing the picture to jump about. See also *Camera moves, Truck*. See Figure 1–40.

Stock footage   A videotape library of shots that are available for use in videos; for example, exteriors of buildings, spectacular sunsets, or a raging fire, all visual elements that could add to the believability of a program.

Subtitle   A secondary title to the program, for example, the date or someone's name.

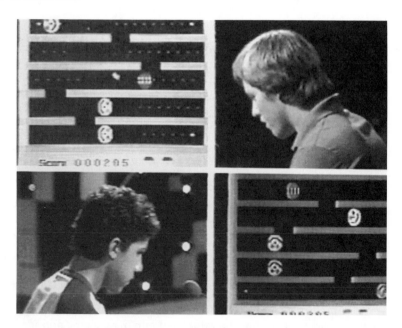

Figure 1-39   A split screen using more than one picture can be used for effect.

Figure 1-40   A Steadicam JR®
camcorder stabilizing system provides
ease of movement without causing a
jumpy picture. (Photo by Joe Lipton.
Courtesy of Cinema Products.)

Super Superimposing one picture or graphic over another, for example, to super a title over an action sequence to begin the program. Also called a *key*. See Figure 1–41.

Supply reel The reel that is supplying the tape for play forward or the reel on which the tape rewinds.

Sweeten See audio sweetening.

Switcher A computerized machine that is designed to connect various sources, for example, recorders and cameras, allowing you to switch between these sources. A switcher may also allow you to mix the video sources, to do dissolves, to add graphics or sound live by microphone, or to do special effects, such as wipes. See Figure 1–42.

(a)

(b)

(c)

Figure 1-41 A title can be supered over an opening scene.

Figure 1-42 A switcher is designed to connect various sources, allowing you to switch between these sources. It may also be able to do dissolves, add graphics or sound, or do special effects. Pictured is Shomi's Adventurer Switcher with a Cheetah TBC (time base corrector) attached. (Photo courtesy of Shomi Corporation.)

Sync Short for synchronized, synchronization, or synchronous; this refers to the speed or speed relationship between machines; for example, the videotape longitudinal speed must be in sync with the playback head or the tape will not play. Independent video sources, like a VCR and a camera, must be synchronized to each other to be able to talk without causing a glitch in the picture. The camera says, "Here's the shot." Sync also refers to the various components of the video signal being in time with another.

Synchro editing A method of synchronizing the play and record machine in an editing system.

Take Accompanied by a number, a take is the designation given to a shot; for example, take 3 means that this is the third time you have recorded the shot. Take is also used to indicate which shot is the best; for example, "that's a take" means that's the one I want when I go to the edit, so make a note.

Takeup reel The reel on which the videotape is wound when in the play or fast-forward modes.

Tease A sequence designed to get the viewer interested in watching the program. It may be a montage of scenes from the program or it may be the beginning of the program.

Time base corrector An electronic device used to correct video signal instability by replacing the sync during playback of a recorded videotape when dubbing. TBCs are used to put the timing of multiple video sources in sync with each other when they are combined on one recording, such as dissolves, wipes etc. Also known as a TBC. See Figure 1–43.

Time code A method of indexing videotape by giving every frame an hour, minute, second and frame designation. Time code is recorded directly onto the videotape without interference with the video, either on one of the audio tracks (Longitudinal Time Code) or within the vertical blanking interval on the tape (Vertical Interval Time Code). Time code counts in real time as the tape runs through the machine, recording hours, minutes, seconds and frames (30 frames to a second by NTSC standard). Time Code editing is more accurate than control track editing because time code is recorded directly onto the tape and will not change. Control track pulses may read as time code numbers but these numbers are not recorded on the tape but rather are counted electronically, one pulse for every frame, during tape playback. SMPTE (Society of Motion Picture and Television Engineers) time code is the world standard. See Figure 1–44.

Timing sheet A sheet of paper that indicates the length of the program by acts, if it is divided into acts, and the length of any interruptions in the program, such as those for commercials. See Figure 1–45.

Title The name given to the program; it may appear at the very beginning as first video or it may come later after the story has already begun.

Tone An even, standardized 10- to 60-second audio signal of 1,000 megahertz recorded at the beginning of the master material or edit master and used to set audio levels during playback.

Tracking Monitoring the speed and angle at which the tape is passing over the video heads. If a tape does not track right, it is not lining up the frame playback correctly and thus scan lines or breakup may appear on the picture.

Figure 1-43 A time base corrector (TBC) is an electronic device used to correct video signal instabilities, putting the timing of multiple sources in sync with each other. Pictured is Shomi's Cheetah TBC. (Photo courtesy of Shomi Corporation.)

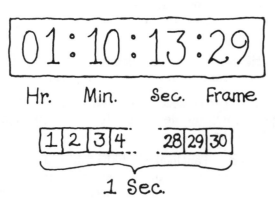

Figure 1-44 SMPTE time code reads hours, minutes, seconds, and frames. It takes 30 frames to make 1 second of video by NTSC standards. (James R. Caruso/Mavis E. Arthur, A BEGINNER'S GUIDE TO PRODUCING TV© 1990, p. 15. Reprinted by permission of Prentice Hall, Englewood Cliffs, New Jersey.)

TIMING SHEET AND CERTIFICATE OF PERFORMANCE

AIR DATE _____   THE VIDEO GAME   STATION _____

SHOW NUMBER   003   DROP FRAME _____   TIME 29:58 INC. COMMERCIALS
                      NON-DROP FRAME XXXX

		HOURS	MIN	SEC	FRAMES	COMMERCIAL NAME/CODE	NOTE
COMMERCIAL 1		01	00	00	00	DR. PEPPER JUNGLEMAN	
ACT I	IN	01	00	30	00	Includes :15 Billboard	
	OUT	01	07	23	00		
COMMERCIAL 2		01	07	24	00	SEGA "FOREIGNER"	
3		01	07	54	00	HERSHEY "COUSIN WILLIE"	
4		01	08	24	00	BLUTH ANIMATION	
5		01	08	54	00	KIDS FOR KIDS "OWENS"	
ACT II	IN	01	09	25	00		
	OUT	01	18	20	00		
COMMERCIAL 6		01	18	21	00	DISNEY "RAINBOW BRIGHT"	
7		01	18	51	00	" " " "	
8		01	19	21	00	" " " "	
9		01	19	51	00	" " " "	
ACT III	IN	01	20	22	00		
	OUT	01	23	58	00		
COMMERCIAL 10		01	23	59	00	SEGA "A FUNNY THING"	
11		01	24	29	00	HERSHEY "COUSIN WILLIE"	
12		01	24	59	00	MICHAEL JACKSON PSA	
13	xxxxxxxxxxxxxxxxxxxxxxxxxxxxxxxxxxxxxxxxxxxxxxxxxxxxxxxxxxxxxxxxxxxxx						
ACT IV	IN	01	25	30	00		
	OUT	01	29	08	00	Includes :15 Billboard	
COMMERCIAL 14		01	29	09	00	HERSHEY "COUSIN WILLIE"	
TO BLACK		01	29	39	00		

Figure 1–45 A timing sheet indicates the length of the program by acts and the length of any interruptions in the program.

Transition   The method used to go from one shot or scene to the next. Some common transitions are the following:

Cut   An abrupt change from one shot or audio to the next. This is the most common transition; it is also the easiest to do, especially if both audio and video are cut simultaneously.

Fade   The gradual increase or decrease of the video image or the audio. Video images generally fade in or fade out from or to black or another color, while audio fades up or down. Video and audio may cross fade or segue, meaning one may come up while another goes down. Most videos begin with a fade in from black and end with a fade out to black.

Dissolve   To gradually fade out one image while simultaneously fading in another one. A dissolve may be used to show the passage of time or to soften the transition between one shot and the next. A dissolve requires two playback machines, one designated the A machine, the other the B machine. The primary source material (raw footage), the A roll, is played back in the A machine. The B roll is played back in the B machine. The B roll may include shots from the A roll dubbed over specifically for the purpose of doing the dissolve. The dissolve can go from A to B or B to A and is recorded on the edit master, the third machine.

Wipe   To go from one shot to another using a geometric pattern, for example, a straight line up, down, or across; or a circle in or out. A wipe literally wipes away one shot while bringing on a second shot or a color background. A soft wipe fuzzes the edges of the wipe, making them somewhat translucent, rather than hard edged.

Jump cut   Moving a person or object from one location to another without giving the viewer any clues as to how the person or object managed to move; for example, a person is inside a house and the scene cuts to outside with the same person now getting into a car. A jump cut tends to be jarring to the viewer. If not used for effect, it can be perceived as a mistake or a lapse in continuity.

Transition rate   The amount of time it takes to make a transition; that is, to dissolve from one picture to the other, the transition rate might be 45 frames. Thus from the time the dissolve begins until the time the dissolve ends, 45 frames pass.

Trim   Changing an edit-in or edit-out point to trim off and add on, usually in frames. An edit in can be trimmed two frames for instance.

TV time   Condensed real time to move the story along. See also *Real time*.

Underscan monitor   A display feature on a monitor that shows the complete video frame, right to the black edges. An editor or engineer uses this to see edge problems that are affecting the video so that they can make adjustments to correct them.

V edit   An edit that records video only (see also *VOS*).

VCR (video cassette recorder)   A machine designed to record and playback video and audio recorded on tape wound inside a cassette box (see also *VTR*). See Figure 1–46.

Vectorscope   Electronic equipment used to set up the color signal on the videotape. Colors are displayed as electronic signals, making possible more precise adjustment than is possible by just looking at the color. See Figure 1–47.

Vertical blanking interval   The period when the picture tube goes blank between the end of the scan of one field and the beginning of the scan of the next field. The vertical blanking interval is sometimes used to record time code (VITC), captioning or test and alignment signals. See Figure 1–48.

Vertical interval time code (VITC)   Time code that records in the vertical blanking interval in the storage spaces there. See also longitudinal time code.

Video   The visual portion of the video recording.

Video signal   All the video information necessary to broadcast or record.

Video source   Any camera, recorded videotape, or other device that generates a video signal.

Videotape sizes   The size of the videotape designated by its width or the recording method; for example, 8mm, Hi 8, $\frac{1}{2}$-inch VHS, $\frac{1}{2}$-inch S-VHS, $\frac{1}{2}$-inch Beta, $\frac{3}{4}$-inch, 1-inch, or 2-inch quad. See Figure 1–49.

Voice over (VO)   Talking over the video without ever seeing the person speaking; it can be used to describe or clarify what is taking place on the video (for example, narration). Voice over may be edited first or last on an edit master. If it is recorded first, as it sometimes is when editing a 30-second TV commercial, the video may then be edited to what the voice over says; for example, in a commercial for a furniture sale, when the voice over talks about sofas, the video would cut to sofas.

Volume unit meter   A volume unit meter (or VU meter) displays the strength of the audio signal with a needle or LED indicator that moves up and down as the sound does.

VOS   Video Over Sound records video only, while leaving the sound already recorded intact.

VTR (videotape recorder)   A machine designed to record and play back video

(a)

(b)

(c)

(d)

Figure 1-46 VCRs designed to record and playback a video signal from tape wound inside a cassette come in a variety of tape playback styles and tape sizes, including $\frac{1}{2}$ inch, 8mm, and ¾ inch. Pictured are four different styles designed to playback and record the VHS $\frac{1}{2}$-inch format. (Photos courtesy of Hitachi and Panasonic Audio/Video Systems Group.)

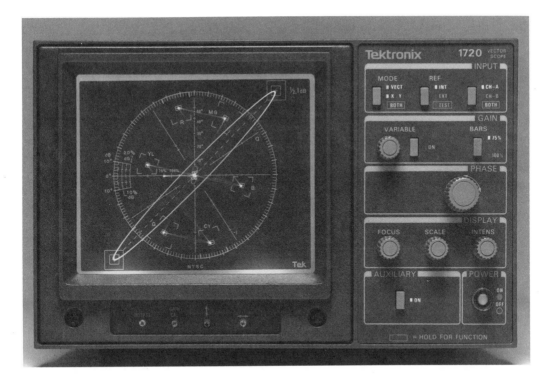

Figure 1-47 A vectorscope displays the video signal as electronic signals for precise color adjustment. (Photo courtesy of Tektronix, Inc.)

and audio on tape wound around an open reel (see also *VCR*). See Figure 1-50.

VU meter See *Volume unit meter.*

Waveform monitor Test equipment used to display and analyze video signal information. See Figure 1-51.

White balance To adjust the color based on what white looks like in a particular location given the lighting conditions. In other words, by showing the camera what white looks like, it automatically knows what all the colors look like and adjusts its electronics accordingly.

Window dub A duplicate of the recorded material with time code burned into the picture; the time code is actually recorded over the picture. A window dub is used to make an accurate edit list, including time code numbers, of the desired shots prior to going into the edit session. A window dub protects the original material from damage from repeated plays during the shot-selection process. See Figure 1-52.

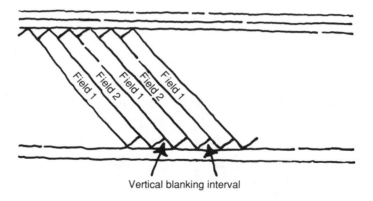

Vertical blanking interval

View from Recording Side of Video Tape

Figure 1-48 The vertical blanking interval occurs between the end of the scan of one field and the beginning of the scan of the next field.

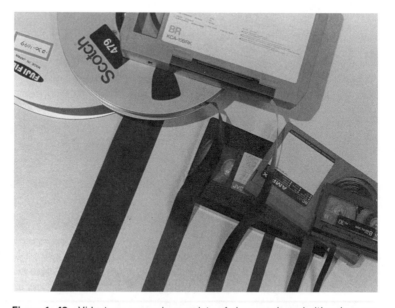

Figure 1-49 Videotape comes in a variety of sizes, packaged either in a cassette or wound around an open reel. (James R. Caruso/Mavis E. Arthur, A BEGINNER'S GUIDE TO PRODUCING TV© 1990, p. 4. Reprinted by permission of Prentice Hall, Englewood Cliffs, New Jersey.)

Figure 1-50 A VTR (videotape recorder) is a machine designed to record and playback video and audio on tape wound around an open reel. Pictured here is a 1-inch VTR designed for 1-inch tape playback and record. (Photo courtesy of Ampex Corporation.)

Working dub A duplicate of the recorded material created for the purpose of repeated viewing and/or making edit decisions.

Work tape A scratch tape on which to practice edits, make a B roll copy, dub over audio or create titles, and the like.

CONCLUSION

Video has its own language. The preceding glossary provides just some of the terms used in video editing and post production. Refer back to them as you read on.

Figure 1-51 A waveform monitor displays the video signal so that it can be analyzed. (Photo courtesy of Tektronix, Inc.)

Figure 1-52 Time code is burned into the picture on a window dub for the purpose of making edit decisions.

TO DO

Can you talk video? Test yourself with the following sentences. Replace all underlined copy with the correct video language.

1. From black, go to on a shot across the street of the Court House. Ten
          (1)   (2)

picture scans into the shot, start judge talking. Take out the Court House shot as
         (3)   (4)   (5)

you bring in a face shot of the judge continuing to talk. Then go to a waist shot of
         (6)   (7)   (8)   (9)

the court reporter keeping a record. Add the sound of a gavel rapping on a desk
                       (10)

and make the crowd sound bigger.
       (11)

2. Make a copy of what you shot and what you edited.
       (12)   (13)   (14)

3. U.S. standard color video signal records 30 pictures per second. Every
         (15)   (16)

picture is created by interlacing two half-pictures.
               (17)

Answers

1. Fade up on EWS of the Court House. Ten frames in, start judge voice
    (1)   (2)   (3)   (4)

over. Dissolve to a CU of the judge on camera. Cut to MS of the court reporter
(5)   (6)   (7)   (8) (9)

keeping record. SFX: gavel rapping on desk. SFX: sweeten crowd.
       (10)   (11)

2. Make a dub of original footage (or raw footage) and edit master.
     (12)   (13)   (14)

3. NTSC records 30 frames per second. Every picture is created by put-
   (15)   (16)

ting together two fields.
     (17)

2

How TV Works

Television came first. It was years before videotape was developed, so most early television shows as well as the first commercials were live and uncut. Whatever happened, happened. There was no way to change it because there was no way to record it unless you were prepared to go to the expense of shooting and editing on film. Most were not.

The beginnings of television are somewhat vague. Sources disagree about when certain developments occurred or whether they were even significant to the development of the medium. It is generally agreed, however, that it all started as far back as the 1800s. In the mid-1800s, Alexander Bain, a Scottish watchmaker, applied for a patent for what he called an "automatic telegraph," which was able to transmit pictures by wire using electronic pulses. The first real demonstration of television came in the early 1900s when John Logie Baird of Great Britain and an American, C. Francis Jenkins, simultaneously and independently, developed a spinning disc system that transmitted fuzzy shapes.

Vladimir K. Zworykin developed a basic television camera around the same time, and in 1927 Philo T. Farnsworth of the United States departed from the mechanical spinning disc approach and took a giant step by developing an electronic method of transmitting a picture. It was not until 1939, however, that television began to exhibit its true potential as a medium for the masses. That was the year that a fledgling National Broadcasting Company (NBC) began transmitting the first regular broadcast shows. It was the beginning of "something big," television as we know it today.

A television signal could be sent, but there was still no way to easily record it. If a broadcaster wanted to delay airing a program, a consideration mainly for West Coast broadcasters since most programming originated on the East Coast at the time, the program had to be recorded by setting up a film camera in front of the TV set and filming it. This was an expensive and technically inferior process called kinescoping.

Some say this was the golden years of television when most programs were real, not manufactured. Television was essentially a live medium. It was the years of such shows as "Ed Sullivan's Toast of the Town," "Arthur Godfrey's Talent Scouts," "Cavalcade of Stars," "Your Show of Shows," "I Love Lucy," "The Red Skeleton Show," "You Bet Your Life," "The Jack Benny Program," and "The Jackie Gleason Show."

Television began to change in 1956 when the first videotape recorder, which used 2-inch quad tape, was introduced by Ampex Corp. See Figure 2-1. In 1958 the first videotape editing was introduced. See Figure 2-2. The development of videotape and videotape editing revolutionized television. Videotape made it possible to record programs easily and cheaply, *and* the picture quality was much better than kinescoping since the signal could be recorded directly.

In its early days, videotape editing was a mirror copy of film editing with editors cutting and splicing tape. See Figure 2-3. It was not until the early 1960s that electronic editing became a reality. In principle, this early electronic editing was the same as it is today; it required an editor to select shots and then put them in the desired sequence by recording them a second time on a second piece of tape, thus creating an edit master.

Electronic videotape editing had some major advantages over film-type, cut-and-splice video editing. Damage caused by handling the tape was reduced. Shots could be used and reused since the original recording remained intact. And the cost of recording and editing programs was considerably lower. Even

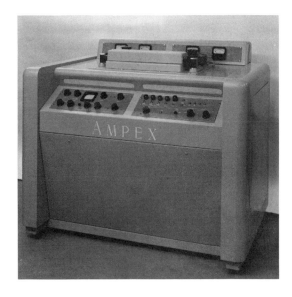

Figure 2-1 The first videotape machine, the VR-1000 quad VTR, was designed by Ampex Corporation; it played and recorded on 2-inch tape. (Photo courtesy of Ampex Corporation.)

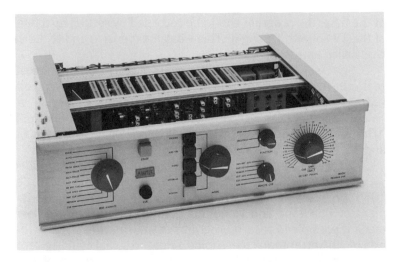

Figure 2-2   Ampex introduced Editec, the first electronic videotape edit system that did not require cutting the videotape. (Photo courtesy of Ampex Corporation.)

with all these advantages, the electronic editing of the 1960s did not compare with today's editing capabilities. There was nothing fine-tuned about the first electronic videotape editing. Computers did not control the edit-in and edit-out points as they do today. Editors pushed the record button when the desired shot was visually seen. The method was close, but not accurate, and was referred to

Figure 2-3   In its early days, videotape was edited by cutting it like film and then splicing the desired shots back together. Pictured is one of these early videotape cutting blocks.

by some as punch and crunch editing, a term still used today to describe basic editing. Loosely, it means you punch the button and crunch the program together.

Punch and crunch editing can be found today on some basic home editing systems and even in some professional edit houses, where it is usually used to make a rough cut. Basically, all you need to do punch and crunch editing is two VCR's wired together. It is by no means accurate, but, as they found out in the beginning of videotape editing, it is better than nothing at all. The other side, of course, is the latest high-tech computerized editing, which can do virtually anything, from any kind of transition to wipes to manipulation of the picture, to . . . well, anything.

If you want to be a good editor, you need to know something about both the basic edit as well as the computerized edit. You also need to know how a television set makes pictures, how videotape records, and what time code is. The answer to these questions and more follow as we discover how TV works.

MAKING THE PICTURE

Put succinctly, television uses electronic pulses and electronic interpretation of those pulses to transmit a video signal. The monitor or television set is the receptor to which the converted signal is sent and transformed into a picture. The picture itself is made up of a series of lines, displayed one line at a time. NTSC, the color video system standard in the United States, displays 525 lines of information to create just one complete picture. The electronic beam creating the lines begins in the upper-left corner of the picture tube and, moving from left to right, scans down the television tube, creating the odd-numbered lines in the picture first, lines 1 to 525. Then the electron beam is told to return to the top center of the television tube and scan down again, interlacing the even-numbered lines, lines 2 to 524, with the odd-numbered lines. Half a picture scan (odd lines 1 to 525 or even lines 2 to 524) is called a field, and one complete picture (two fields) is called a frame. See Figure 2–4. In the United States the standard is 30 frames to make 1 second of video (NTSC).

Since it takes only $\frac{1}{60}$ second to create one field of video and a complete picture is created in $\frac{1}{30}$ second, the eye cannot see one field as it is built line by line and most cannot see a completed frame. However, experienced editors and others who work with videotape all the time *can* see one frame, particularly if it has a glitch or some other problem.

A frame of video is a combination of electronic pulses and sine waves called the composite video signal, that together carries the information necessary to create television. A composite video signal includes seven pieces of vital information:

Figure 2-4 Field 1 scans the odd lines, 1 to 525. Field 2 scans the even lines, 2 to 524. Two complete fields make one frame of video (one complete picture). Note on the picture that the lines between the lighted boxes are broken. In actuality, they are solid. What seems to be a break in the line is happening because only one field has been scanned. Therefore, only half the picture is there.

1. **Horizontal line synchronizing pulses** (horizontal sync): tells the scanning electron beam to start all fields of incoming information at the same place every time.
2. **Picture luminance:** sets the brightness levels of the scanned image.
3. **Color saturation:** determines the color intensity of the incoming signal; the greater the saturation, the deeper the color.
4. **Color reference burst:** tells the monitor or television set to duplicate the colors just the way they look on the recorded tape.
5. **Color hue:** sets the shade of the color on the monitor or television set.
6. **Reference black level:** tells the monitor or television set what video black looks like.
7. **Vertical synchronizing pulses** (vertical sync): controls the vertical blanking interval.

What Are Blanking Intervals?

Within the composite video signal are two other important elements where no picture information is being transmitted. These are the (1) horizontal blanking interval and (2) vertical blanking interval.

The horizontal blanking interval begins at the end of a completed frame of video (the end of the second field) and ends at the beginning of the next frame (the beginning of the first field of the next frame). The horizontal blanking interval occurs as the electron beam returns to the upper left of the picture tube to begin the first field of a new frame. Within this interval, the picture tube goes blank while the color reference burst for the incoming frame is transmitted. See Figure 2–5.

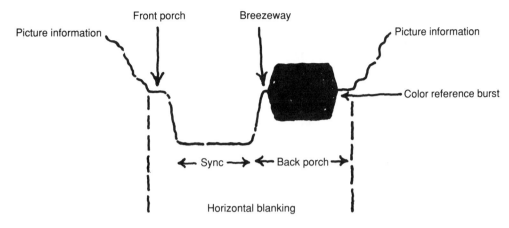

Figure 2-5   The horizontal blanking interval occurs between fields.

The vertical blanking interval begins at the end of field 1 and ends at the beginning of field 2. Within the vertical blanking interval are 20 lines of blanking. Think of these as 20 storage spaces for information. Six are devoted to equalizing pulses that align field 1 with field 2 so that they interlace properly to create the complete picture. Three others store vertical synchronizing pulses that tell the electron beam where to begin displaying the next field. The remaining storage spaces can be used to record the time code (VITC), captioning, or test and alignment signals. See Figure 2–6.

NTSC, PAL, and SECAM

There are three world standards for transmitting a color video signal: NTSC, PAL, and SECAM. NTSC (National Television Standards Committee) is used in the United States, Canada, Chile, and Japan and scans 525 lines in every frame with 30 frames every second. The other two, PAL (Phase Alternating Line) and SECAM (Séquential Colour á Mémoire), scan 625 lines per frame with 25 frames every second. PAL is the standard in more countries than either NTSC or SECAM as noted in the glossary.

There are enough differences between these three standards so that a videotape recorded using PAL will not play on a VCR set up for NTSC or SECAM, and vice versa. In fact, none of the systems is compatible. If necessary, however, PAL video can be converted to NTSC or SECAM to PAL or whatever, but at considerable expense. For the most part, the differences are not a problem since there is not much mixing of the standards. It would be rare to find both PAL and NTSC recorded tape, for example, in an edit. Generally, U.S. programs are recorded on NTSC, Italian programs on PAL, Soviet programs on SECAM, and so on.

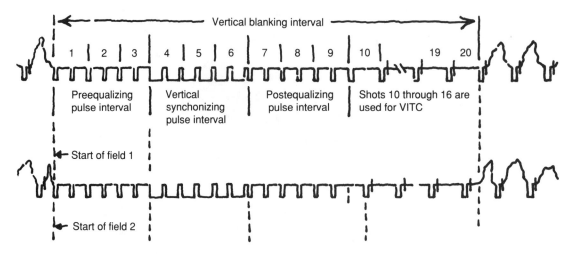

Figure 2-6 The vertical blanking interval provides a storage area for video alignment information and for time code, captioning, or test and alignment signals.

THE RECORDING PROCESS

Videotape is not like film. You cannot see the pictures on videotape except through playback. Hold a piece of videotape up to the light and you cannot tell whether there are pictures recorded on it or not. This can be a problem. The worst of these problems is that original material can be recorded over by mistake. On the positive side, videotape does not have to be developed before it can be viewed like film. You can see videotape immediately and at no added cost. Also, videotape, unlike film, can be erased and reused.

Videotape is categorized by its width and comes on a reel or packed inside a cassette. Five different sizes are available.

1. **2-inch quad, reel to reel:** 2-inch quad was the first videotape accepted as standard by the broadcast industry. Although there are still some 2-inch VTRs around in some television stations and post-production houses, 2-inch quad playback has generally been phased out.
2. **1 inch, reel to reel:** Used in broadcast and some industrial productions and by TV stations.
3. **$\frac{3}{4}$ inch, cassette:** Used in broadcast and industrial productions and by television stations.
4. **$\frac{1}{2}$ inch, cassette:** Used in consumer, industrial, broadcast and TV station productions.

5. 8mm, cassette: Used by consumers mainly, but gaining some inroads into industrial and TV station productions.

While the sizes are different, videotape does not change from the largest to the smallest in terms of its basic construction and how it records pictures and sound. Videotape consists of a very thin layer of oxide on a Mylar-based film backing. As it passes over the video recording heads, the oxide on the tape is magnetized. Later the video heads will decipher these magnetic impulses and play them back as pictures and sound. See Figure 2–7.

Because the magnetic coating is so thin, videotape can be easily damaged by excessive handling, by holding it against the heads in the pause mode too long, or by storing it improperly. Videotape stored on a reel, such as 1- or 2-inch tape, is particularly susceptible to edge damage, meaning the sides of the tape may be damaged. Cassette-stored tape is more likely to be damaged when it is held in pause too long. To compensate for these problems, 1- and 2-inch tape are generally stored on metal reels in hard plastic boxes, and most $\frac{1}{2}$-inch VCRs have a built-in pause release after a certain amount of time passes.

You cannot buy color videotape or black videotape because all videotape is designed to record both. Black and white and color are a function of the picture you feed the tape, not of the tape itself. There used to be black and white cameras, but there are not many of these around anymore except for surveillance and graphic cameras. Today, black and white is created by turning off the chroma, usually during playback during the edit.

A videotape consists of a control track, a video track, and one or more audio tracks. See Figure 2–8.

- The control track is a series of electronic pulses, one for every frame, occurring at the beginning of every frame recorded. It stabilizes the longitudinal motion of the tape during playback, somewhat like sprocket holes on film.

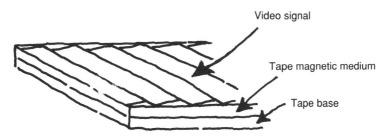

Figure 2–7   Videotape consists of a very thin layer of oxide on a Mylar-based film backing. The oxide is magnetized as it passes over the record heads.

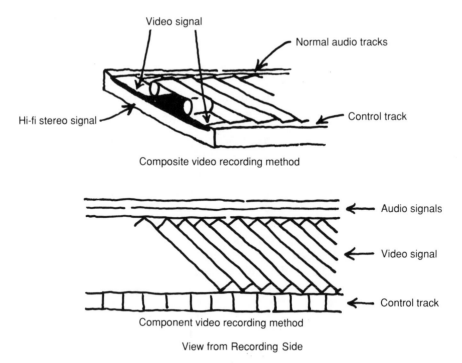

Figure 2-8 shows:
- Video signal
- Normal audio tracks
- Hi-fi stereo signal
- Control track

Composite video recording method

- Audio signals
- Video signal
- Control track

Component video recording method

View from Recording Side

Figure 2-8 A videotape recording consists of control track, a video track, and one or more audio tracks.

Control track can also be used for a type of computerized editing when time code is not available.

- The video track carries the composite or component video signal, which includes all the picture information and horizontal and vertical sync pulses.
- One to four audio tracks are available. Typically, consumer equipment will record and play back one to two longitudinal tracks, while broadcast equipment will record and play back up to four. One of the four tracks is sometimes reserved for recording time code. The original material audio may be recorded on one track while the second is reserved for music or sound effects or additional voice over. Audio tracks are generally kept separate on the edit master and mixed only when making distribution dubs.

What Is Time Code?

First, time code is not the same thing as control track. Control track is on every recorded tape and is a series of pulses, one for every recorded frame. Some editing systems will read control track in time code-like numbers, noting hours, minutes, seconds, and frames, *but* the system is really only reading the pulses and moving the counter one frame for every pulse. It is not reading actual recorded time code numbers. Control track changes when you zero the counter or pop the tape. It can also slip several frames in its count with repeated shuttling of the tape. Thus control track is a function of the machine's mechanical count system and is not as accurate as recorded time code.

Time code records numbers (hours, minutes, seconds, and frames) as electronic pulses directly onto videotape. The world standard time code is SMPTE (Society of Motion Picture and Television Engineers) time code. Time code recorded on a tape does not change from playback to playback or machine to machine. If a shot is located at 00:22:13:07, meaning this shot can be found at 0 hours, 22 minutes, 13 seconds, and 7 frames on the tape, it will always be at this location.

Time code can be longitudinal (LTC) *or* it can be vertical interval time code (VITC). See Figure 2–9. Both are essentially the same except for these differences:

1. LTC records on an audio track.
2. VITC records in the vertical blanking interval, the area on the tape between the end of field 2 of a frame and the beginning of field 1 of the next frame.
3. VITC requires no special amplification or signal-processing equipment during playback. LTC does.
4. VITC has a code that protects against read errors. LTC does not.

Both VITC and LTC have some problems. LTC may misread when the tape is moving very slowly or may not read in still frame. VITC may not be readable at fast speeds. As a result, to be safe, editors may elect to record both, one mirroring the other.

It is best to record time code sequentially from the beginning of a tape to the end with no break in code. If the time code changes, for example, 01:00:20:15 jumps to 01:03:25:30, it can take more time to find a shot. The jump will disorient the editor and can cause the computer edit system to drag as it searches for the numbers.

Why would time code jump like this? It probably never will on the edit master, but it can on the original material. If the time code generator on the record VTR or VCR is counting in real time, for example, it keeps running between shots just like the clock. Say you end a shot at 01:00:35:42 and stop tape. When you

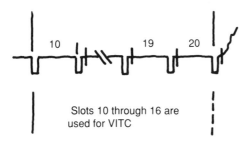

Slots 10 through 16 are used for VITC

Figure 2-9   Vertical interval time code (VITC) is stored in the vertical blanking interval.

begin recording again, it may be 10 minutes later and the next shot may begin at 01:10:35:42. The time code jumps from shot 1 to shot 2. This is not a huge problem, but it may be irritating during the edit, especially if you do not know it is going to happen. Just keep it in mind when you are recording and when you are editing.

Nondrop versus Drop Frame Time Code

Not to confuse you, this is *not* another kind of time code. It is rather the difference between TV time and Real time and two different ways to record time code on an edit master. Nondrop frame time code counts a full 30 frames every second. As a result it is longer than real time. Drop frame time code also records hours, minutes, seconds, and frames, but does it irregularly to make it coincident with real time. That is, it drops or skips 108 frames every hour. In other words, the time code jumps two frames, every minute except on the tenth minute. This is done to synchronize recorded time with real time.

This was necessary when color TV was introduced along with NTSC. The FCC wanted to make sure that all TVs, both color and B/W, could receive the same broadcast signal. The B/W broadcast system was synced to 60 cycles (60 hertz) per second. When the chroma signal was combined with the luminance signal, it crosstalked with the audio on home TVs causing distortion. The engineers found that by reducing the sync to 59.94 cycles per second (59.94 hertz) the crosstalk problem was eliminated and B/W TVs were still compatible with the color broadcast system.

The NTSC television signal operates at 59.94 hertz. This is TV time. A clock or real time operates at 60 hertz. As a result, TV time is slightly behind real-time time, 108 frames per hour to be exact. Thus, if a half-hour program is recorded using drop frame time code, the edit master will run in real time and end a half-hour. If it is recorded using nondrop frame time code, it will run in TV time and end 54 frames later (approximately 2 seconds too long).

The difference between drop frame and nondrop frame time code is critical only if the program must fit into a certain time period, like those airing on broadcast television. Even so, both are used in editing broadcast programs; but those using nondrop frame time code are shorter to accommodate the difference. That is, a half-hour program will be edited to 22:37 in length with nondrop frame time code, but edited to 22:39 with drop frame time code. Both programs will actually run the same length in real time when they air, 22 minutes, 39 seconds.

PUTTING IT ON TAPE

Knowing the basics of how TV works includes understanding some techniques of getting it on tape the way you want it. Anyone can record videotape, but not everyone can do it well. Learning all you can about the creative side as well as the electronic side will help assure that you are among the few who can do it well.

Television has four basic ways to frame a shot, three basic camera angles, and six ways a camera can move on the action. It is from these basic elements that camerawork is planned.

Basic Camera Framing

The framing of a shot is designated based on the size of the main subject matter in relationship to the picture itself, as follows:

1. **Extreme wide shot (EWS)** (see Figure 2–10): This is a wide, sometimes panoramic shot of a location or action, used generally to establish location, for example, the exterior of the hospital where our actor is being operated on. This shot has minimal use in terms of forwarding the story line due to the fact that the TV screen where the viewer sees it is small in contrast to its view of the scene.
2. **Wide shot (WS)** (Figure 2–11): Full action or location with some distinguishing characters or action visible. A WS might be used to open a scene, for example, in a living room, to establish where everyone is seated or standing and what the room itself looks like. It may also be used for some of the action or interaction between the characters.
3. **Medium shot (MS)** (Figure 2–12): Generally, a waist up shot of one of the actors, but it could be a halfway shot of action or an inanimate object. The viewer gets a closer look with this shot, and it also allows the actor to dis-

Figure 2-10 An extreme wide shot (EWS) establishes location, for example, a train yard.

Figure 2-11 A wide shot (WS) may be used to establish location, preempting the need for an EWS.

play emotions to the viewer. A MS may include more than one character, for instance, a medium two shot.

4. **Closeup (CU)** (Figure 2–13): An even closer shot, full face or bust up or even closer. This type of shot puts the viewer in direct contact with the actor or event, since they are able to see everything in detail. A variation of this is an ECU, or extreme closeup. See Figure 2–14.

Given the size of the TV screen, the most effective shots are the MS, the CU, and variations on the WS.

Figure 2-12 A medium shot (MS) gives the viewer a closer look at the subject.

Figure 2-13 A closeup (CU) gives detail to a particular part of the picture.

Basic Camera Angles

In addition to framing, the angle of the camera is critical to the shot. The camera angle can be high and shooting down, making the performer or action appear small and insignificant. See Figure 2-15. The camera angle can be low and shooting up, making the performer or action seem huge and overbearing. See Figure 2-16. Or it can be straight forward or head on from a normal perspective. See Figure 2-17.

1. **First person or POV** (Figure 2-18): The viewer is put directly into the shoes of one of the characters, with the camera taking that character's position and allowing the viewer to step in visually for example, an astromer looking through a telescope. If there are other actors in the scene, they register

Figure 2-14 An extreme closeup (ECU) goes even closer to the subject.

Figure 2-15 The camera angle can be high, shooting down, making the performer appear small and insignificant.

direct eye contact with the camera, further heightening the sense that the viewer is really in the shoes of the actor.

2. **Second person** (Figure 2–19): The viewer stands alongside the actor and sees from the actor's point of view but not as if in his shoes for example, next to the telescope but not looking through it. No direct eye contact is made with the camera, and thus the viewer maintains his or her objectivity.

3. **Third person or objective** (Figure 2–20): The viewer is removed from the action, an observer only, and is thus not attached to any one character,

Figure 2-16 A low camera angle makes the subject appear larger than life, overwhelming and even overbearing.

Figure 2-17 A normal straightforward camera angle puts the subject on
an equal level with the viewer.

making the viewer more objective in terms of the fate of any of the actors,
for example, outside the observatory.

Both the framing and the angle of the shot are critical in terms of the
viewer's participation in the video. Some draw the viewer into the action, others
remove him. You determine which best suits the intent and story line of the video
you are making.

Basic Camera Moves

So you have decided on the framing of the shot and you know what the action of
the scene will be, but have you decided on how the camera will move? This must

Figure 2-18 First person or POV
(point of view) puts the viewer into the
shoes of the actor. If you were looking
at a POV of an astronomer looking
through a telescope, you might see this.

Figure 2-19 Second person puts the viewer alongside the actor, that is, next to the telescope but not looking through it.

be decided as well. Your options are varied. The camera can remain static (still) or it can move from one location to another or its lens can move. Let's take a look at some of the options.

1. **Static:** Neither the camera or its lens moves. All the action takes place in the scene. Statics are used particularly for dialogue-intensive scenes. For example, two people in a heated argument might be shot static.
2. **Zoom:** The camera lens can zoom in or out, meaning that it changes its focal length. The wider the focal length (smaller number) will have more in focus

Figure 2-20 Third person puts the viewer outside the action, seeing it as an observer, that is, outside the observatory. This shot could also be used as an establishing shot, then cutting to inside the observatory.

than the narrower focal length (larger number). This not only changes the size of the main subject in the shot, but also the things that are visible to the viewer. If it zooms in, it focuses the viewer's attention on the principal object in the zoom. If it zooms out, it broadens the scene to include other aspects. A zoom in might be used when a scene begins with a wide shot. For example, the scene begins with a wide shot of a street. The camera might zoom in to focus on a city limits sign to establish location. See Figure 2–21.

3. **Truck:** The entire camera moves, left or right, parallel to a subject. A truck usually follows the action and can be done with the camera on wheels or with the camera mounted on the cameraperson by way of a Steadicam or hand-held by the cameraperson. See Figure 2–22.

4. **Pan:** The camera moves left or right on its pedestal. The camera itself does not change its location; rather, it scans the scene either to the left or to the

(a)

(b)

(c)

Figure 2–21 A zoom focuses the viewers's attention on a particular object in the picture.

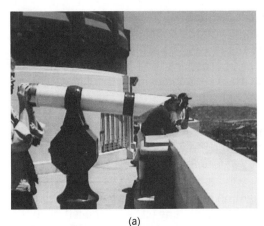

(a)

(b)

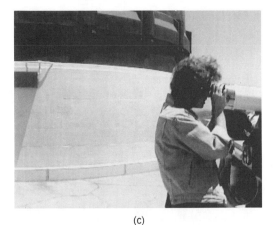

(c)

Figure 2-22 When the camera trucks, the camera physically moves parallel to the subject.

right of its fixed position. A pan might be used to see a scene that will not fit into the framing in its entirety. For example, a pan might be used if it is important for the viewer to know what shops are on a city block. One shot cannot include all of them, so the camera might be positioned in the center of the block and it pans left to right, from one end of the block to the other. See Figure 2-23.

5. **Tilt:** The camera itself does not change its location; rather, it tilts up or down from its fixed position. A tilt might be used to shoot a scene that would otherwise take several static shots to show the action. For example, if a parachutist is coming down, a tilt down would be used to follow him as he descends. See Figure 2-24.

6. **Dolly:** The camera moves closer to or farther away from the main subject in the shot, making it appear larger or smaller. A dolly is different than a zoom

(a)

(b)

(c)

Figure 2-23   When the camera pans, it scans the scene from a fixed position.

in that it does not compress or expand the composition of the shot, as changing the focal length of the lens does. The camera may be mounted on a trucklike device called a dolly and used in a studio location, or it may be mounted on an automobile or other movable object. See Figure 2-25.

The framing, angle, and movement of the camera are determined during production, not during post production. However, it is important to plan these elements of a shot with the post production in mind, because it is there that the editor must make the shots work together. For example, if shot 21 is a pan and shot 22 is a pan also, will these cut together? Does the pan go in the same direction? Do you really want two pans back to back? These are things that should be considered when the shots are being planned.

(a)

(b)

(c)

Figure 2-24 When a camera tilts up or down, it does not change its physical location, but tilts on the tripod head or in the cameraperson's hands.

(a)

(b)

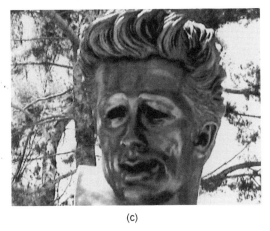

(c)

Figure 2-25 When a camera dollys, it physically moves closer to or farther away from the main subject. Unlike a zoom, a dolly does not compress or expand the composition of the shot. A dolly does not change perspective.

Aspect Ratio

In addition to framing and angles, an editor needs to be aware of television's aspect ratio. Simply, this refers to the ratio between picture height and width and is what the television can see of the scene. Television aspect ratio is 3:4. See Figure 2-26. In other words, for every 3 inches of height, there are 4 inches of width, for example 9 inches by 12 inches. This should not be confused with the size of a television set. Size refers to the diagonal measurement of the tube from upper-left corner to bottom-right corner.

It is important to keep the aspect ratio in mind while planning or framing shots. It is just as important to remember that not all home television sets display the entire 3 by 4 picture. There is something called a *cutoff zone* that may or may not be seen at home, depending on the way the set is adjusted and the

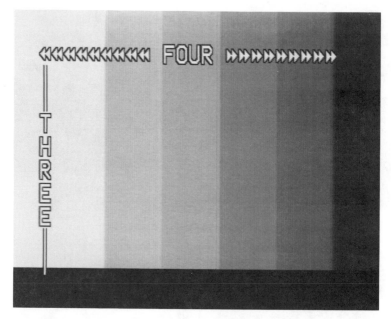

Figure 2-26 TV aspect ratio is 3:4, for example, 9 inches high and 12 inches wide.

transmission is coming into the set. To accommodate for this, there is a safe title area and a safe action area on the 3 by 4 image or scanned picture.

Safe title is the area of the picture that will be readable on home receivers, that is, titles, credits, or any written words. This area is 80% of the total scanned picture. See Figure 2-27.

Safe action is the area of the picture where you are assured that the action will not be cut off on home receivers. See Figure 2-28. This area is 90% of the total scanned picture.

PLANNING A PROGRAM

Television is planning, planning, and still more planning. Plan from the beginning. How good your program finally is will largely depend on how good your plan was and how well you executed it. Begin your plan by looking backward.

Library Tapes

Maintain a video library and an audio library. The video library will include any shots you could conceivably use again. This could include any video shot, but most will be wide, establishing shots. The audio library is a collection of sounds

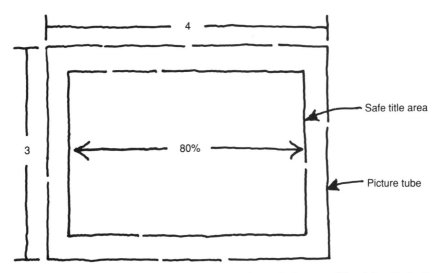

Figure 2-27 Safe title, 80% of the total scanned picture, is the area of the picture that will always be readable on home receivers.

or music that may prove useful in a future production. Both libraries are time savers, and when you save time, you ultimately save money for some other element of the production.

Choose shots or audio for the library from existing material, transferring

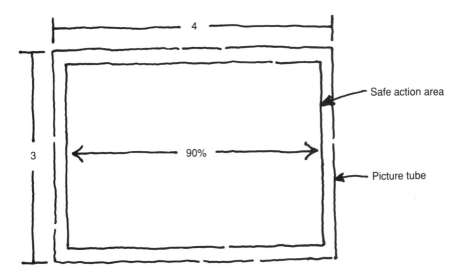

Figure 2-28 Safe action, 90% of the total scanned picture, is the area where action will not be cut off on home receivers.

them over to the library tape; or plan and record specific shots or audio you need or just want to include in the library. In addition to the obvious importance of recording the right shots and audio, the second most important thing in creating a library is devising a workable system so you can find a recorded shot or audio later. Here are a few ideas on how to develop a system of cataloging and maintaining your library:

1. Decide where you will store the library. With a video library, this will be simple. The choice must be videotape. With a sound library, you have the choice of audio or videotape. Both have advantages. If you put it on videotape, it will be easier to log the location of the sounds or music and easier to retrieve them when needed. You could actually put a slate at the location of the sounds so that you could find them visually. You could put audio and video on the same tape so you have only one library, not two separate ones. However, you will tie up a VCR when you are accessing an audio library stored on videotape. That is the biggest disadvantage. Audio tape, on the other hand, will not tie up a VCR but it will be more difficult to locate the desired audio. Weigh the choices and make yours based on your equipment and your own preference.

2. Give every tape in your video and audio library a number. If you only have one tape to begin, it will be tape 1. When you fill up this tape and add a second, it will be tape 2 and so on.

3. Log what is on all tapes. Use the log sheet form provided in the Appendix. Make sure to note not only the video or sound but the number where you can find it. This number will be the counter or the time code number you see when it begins to play on the VCR or the counter number you see when the audio begins to play on the cassette recorder. It will also be useful to note where the sound or visual ends. When you want to find a selected video or audio segment, using the log sheet, insert the cassette on which it is recorded. Rewind to the beginning of the tape and zero the counter. Then roll forward to the number noted on your log sheet. Even if you have access to control track, use this system since control track is mechanical and not recorded. If you have recorded time code on the tape, you will not need to rewind since the location numbers for the wanted video or audio will never change.

4. Make an alphabetical file. Once all video and audio is logged, make an alphabetical list of everything in the library, noting the cassette number and the counter or time code number location. For example:

SUNSET	Tape 2	02:02:34:15
THUNDERSTORM	Tape 1	01:07:45:01
UNDERWATER	Tape 1	01:15:12:05

You could include a more detailed description of the shot, if desired, or anything else that will help you know immediately what is at that particular location on the tape. Put this list on a computer if you have one. If not, put each item on the list on a separate 3 X 5 card. This alphabetical list will be a welcome supplement to the log sheet, making it easier to find a desired video or audio segment.

Remember you can combine your video and audio library, putting sounds on videotape with video shots. For example, the shot of a sunset might be followed by a slate that says "birds singing" with the sound of birds singing recorded. Also, if a video shot has particularly good audio such as the sound of the ocean you might want to note on your log and alphabetical file that it is a source for both video and audio.

Creating a video and audio library will save you time and it will make your videos better because you will have more choices. It will also allow you to reuse videotape. It is smart planning for the serious producer.

Planning Lists

Television runs on paper. As a matter of fact, sometimes it may seem like there is more paper than videotape. Doing all the paperwork may be a drag, but ultimately it will be an asset, for it is through this paper that you will begin to see the program as it will ultimately be.

Putting the show on paper is a big part of the planning process, and reducing the paperwork to lists helps break down the many aspects of a production into a workable form. With comprehensive lists, problems can be kept to a minimum and organization kept to a maximum, and the results will be a number 1 production. The following basic lists will help organize a production from the beginning up to the edit. Where each is used, in preproduction, production, or post production, is indicated.

1. **Master checklist:** For use throughout production. This is a comprehensive list of all elements of the taping. You will use this to make certain that you have completed all aspects of preparing for the taping. See Figure 2–29.

2. **Schedule:** For use throughout production. This is the complete time schedule for the shoot, including preproduction, production, and post production. This schedule will map out the day-to-day activities during the entire production cycle from the initial idea to script, to location scouting, to taping sessions, and to the final edit. See Figure 2–30.

3. **Budget worksheet:** For preproduction use. This worksheet is designed to help estimate the cost of producing the program. See Figure 2–31.

4. **Equipment list:** For preproduction and production use. This is a complete

Show Title __Western Shootout__

Date __12-5__

MASTER CHECKLIST

(Add or delete items as appropriate to your show)

(Check when completed)		(Date completed)
✓	OUTLINE	12-5
✓	BUDGET	12-5
✓	EQUIPMENT LIST	12-5
✓	CREW LIST	12-5
✓	SCENARIO	12-6
✓	FORMAT	12-6
✓	SCRIPT	12-6
✓	STORYBOARD	12-6
✓	CAST LIST	12-6
✓	MAKEUP	12-7
✓	WARDROBE	12-7
✓	PROP LIST	12-7
✓	LOCATION LIST	12-7
✓	SET DESIGN	—
✓	SHOT LIST	12-8
✓	SCHEDULE	12-8
✓	LOCATION SKETCHES	12-7
✓	CAMERA POSITIONS	12-7
✓	LIGHTING PLOT	12-7
✓	BIBLE	12-9
✓	SHOT LOG	12-10+
✓	CREDITS	12-17
✓	EDIT LIST	12-17
✓	DUBS	12-20

Figure 2-29 Master checklist sample. (James R. Caruso/Mavis E. Arthur, A BEGINNER'S GUIDE TO PRODUCING TV© 1990, p. 32. Reprinted by permission of Prentice Hall, Englewood Cliffs, New Jersey.)

SCHEDULE

DATE SCHEDULED	EVENT
12-5	BUDGET COMPLETE
12-5	EQUIPMENT CHOSEN
12-5	CREW CHOSEN
12-6	FINAL SCRIPT APPROVAL
12-6	CASTING
12-7	LOCATIONS SELECTED
12-7	WARDROBE, MAKEUP AND PROPS SELECTED
12-8	PREPARE SHOT LIST
12-8	SCHEDULE TAPING DAYS
12-7	POSITION CAMERAS AND LIGHTING
12-10 thru 12-16	TAPING
12-17	MAKE EDIT DECISIONS
12-18	EDIT
12-20	DISTRIBUTE DUBS

Figure 2–30 Scheule sample. (James R. Caruso/Mavis E. Arthur, A BEGINNER'S GUIDE TO PRODUCING TV© 1990, p. 34. Reprinted by permission of Prentice Hall, Englewood Cliffs, New Jersey.)

list of all the equipment you will need for a professional-caliber shoot, working within the limitations of what is realistically available. See Figure 2–32.

5. **Crew list:** For preproduction and production use. This is a complete list of the crew and their duties. See Figure 2–33.

6. **Cast list:** For preproduction and production use. This is a complete list of all cast members by role and real name. See Figure 2–34.

7. **Location list:** For preproduction and production use. This is a complete list of all locations and the dates and times of the scheduled taping at those locations. This may also include the shot list for the location. See Figure 2–35.

8. **Prop list:** For preproduction and production use. This is a list of all props you will need on the shoot and notes the particular location where they will be used. See Figure 2–36.

9. **Shot list:** For preproduction and production use. This is a complete list of all the shots and any extra sound you want to record on the videotape. See Figure 2–37.

10. **Shot log:** For production and post-production use. This is a list of everything recorded on the videotape, noting its exact location by time code or counter number and any notes on the shot. See Figure 2–38.

Show Title __Western Shootout__

Date __12 - 5__

BUDGET WORKSHEET

Note: Estimate all costs. It is not necessary to make any final decisions about locations, equipment, crew, edit, or set at this time. This budget is a best-guess estimate of all expenses for the proposed video.

VIDEOTAPE

Cost per tape $ 5.98

No. Cassettes   ×   9

                        Total $ 53.82

LOCATION FEE

(Place, if you know)

_____ $ _____

_____   _____

_____

                        Total $ ⊖

TRANSPORTATION

Gas $ 30.00

Parking   15.00

Toll charges   5.00

Airline tickets   ⊖

Misc.   20.00

                        Total $ 70.00

PROPS

Buy $ ⊖

Rent $ 15.00

                        Total $ 15.00

Figure 2-31 Budget worksheet sample. (James R. Caruso/Mavis E. Arthur, A BEGINNER'S GUIDE TO PRODUCING TV© 1990, p. 35. Reprinted by permission of Prentice Hall, Englewood Cliffs, New Jersey.)

EQUIPMENT

(Name it, if possible)

Video _____

_____

_____ $ _____

Audio _____

microphone

20.00

Lighting _____

Pro light

10.00

Misc. _____

_____

Total $ 30.00

CREW

Cost per $ 10.00

No. crew members   × 6

Subtotal   60.00

Plus other   + _____

Total $ 60.00

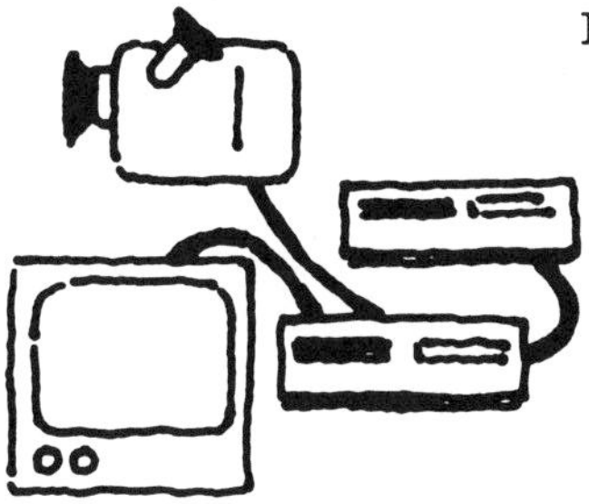

EDIT

Rent/buy equip _____

_____ $ _____

Subtotal $ —0—

Postproduction facility

cost per hour $ 10.00

No. hours est.   × 2

Subtotal $ 20.00

Total $ 20.00

Figure 2-31   (continued)

TELEPHONE

Estimated cost $ _5.00_ $ 5.00

ARTWORK/GRAPHICS

Misc. expenses $10.00

Artist ___—O—___

Total $ _10.00_

CATERING

Estimated cost $ 20.00 $20.00

MUSIC

Buy records/tapes $ 8.00

Original music $ _____

Total $ 8.00

ADDITIONAL VISUALS

Photos $ _____

Video ___3.00___

Film _____

Other _____

Total $ 3.00

POSTAGE/DELIVERY SERVICE

Stamps $ 10.00

Delivery $ _____

Total $ 10.00

PHOTOCOPYING

Estimated cost $ 30.00

Supplies _____

Total $ 30.00

SET

Planning $ ___—O—___

Construction _____

Total $ ___—O—___

Figure 2-31 (continued)

CAST
    Expenses $ 20.00
    Per diem

                      Total $ 20.00

DUB COST
    Cost per tape $ 5.00
    No. cassettes   × 10
    Subtotal $ 50.00
    Dubbing fee $ 5.00
    No. cassettes   × 10
    Subtotal $ 50.00
                      Total $100.00

WARDROBE
    Buy $ —0—
    Rent   30.00
                      Total $ 30.00

MAKEUP/HAIR
    Supplies $ 20.00
    Misc. expense $ —0—
                      Total $ 20.00

TOTAL ALL $504.82

MISC. EXPENSES
    Total all $504.82
    Add 10%   × .10
                      Total $ 50.48

YOUR FEE $ —0—

TOTAL BUDGET $555.30

Figure 2-31 (continued)

Show Title __Western Shootout__

Date __12-5__

EQUIPMENT LIST

CAMERAS

Total no. ___3___

Format (specify number)

VHS __2__   Beta _____   8 MM __1__

Designated duties

#1 __Action - angle #1__

#2 __Action - angle #2__

#3 __Reaction close ups__

#4 _____

MICROPHONES

Total no. ___1___

Type (Specify number of each)

Handheld _____   Boom __1__

Mike stand _____   Lavaliere _____

LIGHTING

Total no. units ___3___

Type (Specify number)

Pro lamps __1__   Shiny board __1__

Camera mounted __1__   Other _____

Figure 2-32 Equipment list sample. (James R. Caruso/Mavis E. Arthur, A BEGINNER'S GUIDE TO PRODUCING TV© 1990, p. 39. Reprinted by permission of Prentice Hall, Englewood Cliffs, New Jersey.)

TAPE

Total no. cassettes ____9_____

Kind (Specify number)

VHS __6__   Beta _____   8 MM __3__

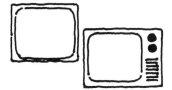

TVs/MONITORS

(Number and size of screen)

No. Black & White _____

No. Color ___2-19" and 5"_____

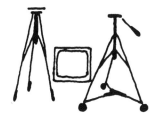

ACCESSORIES

(Specify number)

Tripods/stands

Camera __1__   Lighting __1__

Audio _____   Dollys __1__

Lighting gels (color and number of each)

color __1__   # orange

color __1__   # blue

VCRs

Total no. __1__

Kind (specify number)

VHS __1__   Beta _____   8 MM _____

Figure 2-32   (continued)

Show Title __Western Shootout__

Date __12 - 5__

CREW LIST

DIRECTOR __you__

LIGHTING DIRECTOR __you__

TECHNICAL DIRECTOR _____

AUDIO DIRECTOR __you__

SET DESIGNER _____

SET CONSTRUCTION _____

VIDEOTAPE OPERATOR _____

CAMERA OPERATOR #1 __Tom__

CAMERA OPERATOR #2 __Sandy__

CAMERA OPERATOR #3 __Robert__

CAMERA OPERATOR #4 _____

GAFFER __Edward__

GRIP __Roland__

MAKEUP ARTIST __Michelle__

HAIR STYLIST __Michelle__

WARDROBE __Michelle__

PRODUCTION ASSISTANT __Marie__

ASSISTANT TO:

CAMERA 1 __Ralph__

CAMERA 2 __James__

CAMERA 3 __Melissa__

CAMERA 4 _____

GO-FER __Ronnie__

Figure 2-33 Crew list sample. (James R. Caruso/Mavis E. Arthur, A BEGINNER'S GUIDE TO PRODUCING TV © 1990, p. 41. Reprinted by permission of Prentice Hall, Englewood Cliffs, New Jersey.)

Show Title *Western Shootout*

Date *12 - 6*

CAST LIST

ROLE *Marshal Goodguy*   TALENT *Joe*

ROLE *John Sleeze*   TALENT *Jonathan*

ROLE *Mary Sunshine*   TALENT *Sue*

ROLE *Jerry Lee*   TALENT *Brian*

ROLE *Bartender*   TALENT *Mikael*

ROLE *Sleeze man #1*   TALENT *Jim*

ROLE *Sleeze man #2*   TALENT *Robert*

ROLE *Mayor*   TALENT *Earl*

ROLE *Undertaker*   TALENT *Bob*

EXTRAS *Townsfolk*

Figure 2-34   Cast list sample. (James R. Caruso/Mavis E. Arthur, A BEGINNER'S GUIDE TO PRODUCING TV© 1990, p. 42. Reprinted by permission of Prentice Hall, Englewood Cliffs, New Jersey.)

Show Title __Western Shootout__

Date __12-7__

LOCATION LIST

Locations:
Interior   Exterior

FIELD LOCATIONS	EXTERIORS	INTERIORS
Red Dog Saloon	✓	✓
Streets of San Ramon	✓	
Marshal's Office	✓	✓

STUDIO LOCATION

Director's House		✓

Figure 2-35 Location list sample. (James R. Caruso/Mavis E. Arthur, A BEGINNER'S GUIDE TO PRODUCING TV© 1990, p. 43. Reprinted by permission of Prentice Hall, Englewood Cliffs, New Jersey.)

Show Title _Western Shootout_

Date _12 - 7_

PROP LIST

ITEM	NEEDED AT LOCATION
1 Rifle	Marshal's Office
1 Rifle	Red Dog Saloon
4 Western Gunbelts w/guns	All
Clock	Red Dog Saloon
Clock	Marshal's
Badge	All

Figure 2–36 Prop list sample. (James R. Caruso/Mavis E. Arthur, A BEGINNER'S GUIDE TO PRODUCING TV© 1990, p. 44. Reprinted by permission of Prentice Hall, Englewood Cliffs, New Jersey.)

11. **Edit decision list:** For post-production use. This is your edit plan, a list of everything that will go into the show, in consecutive order. See Figure 2–39.

Since they are a part of the post-production process, how to make and use a shot log and edit decision list is covered more completely in Chapter Four. All these lists and any other material needed to put the program together should be filed together. This can be a loose-leaf file or it can be put into a three-ring binder. In any event, this file becomes the program bible.

Show Title __Western Shootout__

Date __12-8__

SHOT LIST

LOCATION __Red Dog Saloon__

Note: I = Interior   E = Exterior

Framing: CU, MS, WS, ECU, MCU, etc.

I or E	SHOT #	FRAMING	DESCRIPTION
E	22	WS	Saloon
I	23-1	WS	Inside Saloon
I	23-2	CU	John Sleeze
I	23-3	CU	Bartender
I	23-4	M 25	Two men at table
I	23-5	MW	Sleeze at bar

CALL TIME __9 AM__

DAY __December 10__

CALLED CREW (if all, just say "ALL")

__All__

CALLED CAST (if all, just say "ALL")

__John Sleeze__

__Two Townfolk__

__Bartender__

Figure 2-37 Shot list sample. (James R. Caruso/Mavis E. Arthur, A BEGINNER'S GUIDE TO PRODUCING TV© 1990, p. 45. Reprinted by permission of Prentice Hall, Englewood Cliffs, New Jersey.)

SHOT LOGS: WHO SHOT JACK?

Location: Exterior, Regis Mansion

Tape No.	Shot No.	Take No.	Time code/counter numbers In	Out	Audio	Notes
1	1-1	1	01:03:03:00	01:07:07:12	NO	EWS static
1	1-2	1	01:07:07:12	01:10:00:15	NO	EWS push to front door
1	1-3	1	01:10:00:15	01:20:14:23	NO	EWS push to upper window
1	1-4	1	01:21:15:16	01:23:24:10	NO	EWS truck to upper window
1	1-5	1	01:24:19:11	01:26:19:11	YES	CU tree blowing in wind
1	1-6	1	01:27:21:22	01:31:24:11	NO	WS night sky (lightning effect)
2	4-1	1	02:01:10:18	02:04:10:18	YES	POV person running through woods
2	4-2	1	02:05:00:19	02:08:30:23	YES	MS person running through woods
2	5-1	1	02:17:35:12	02:18:45:23	NO	CU face/obscured by hat
2	5-2	1	02:18:45:23	02:19:15:23	YES	POV person falling in woods
2	7-1	1	02:23:16:28	02:24:44:17	YES	MS person getting up from fall
2	7-2	1	02:24:55:19	02:25:45:13	NO	ECU eyes looking off
2	7-3	1	02:11:53:07	02:15:29:09	YES	WS person running away/woods

Figure 2-38 Shot log sample.

EDIT DECISION LIST
T: 04:11

Scene	Tape No.	Shot No.	Edit Type	Transition	Time code/counter numbers (tape no. is designated hour)		Duration	Description
					In	Out		
1	SL		A	FADE UP				THUNDER STORM under through edit 15
	1	1-1	V	FADE UP	01:03:03:00	01:03:13:00	10:00	EWS static house
	1	1-5	V	DISSOLVE	01:24:19:11	01:24:34:11	15:00	CU tree
	1	1-6	V	CUT	01:27:21:22	01:27:31:22	10:00	WS night sky
		EFX	V	KEY				Lightning over
	SL	SFX	A					Accompanies lightning
	1	1-4	V	CUT	01:21:15:16	01:21:45:16	30:00	EWS/truck window
2/3	SL		A	FADES UNDER				THUNDER STORM
	3	2-1	V	CUT	03:07:37:15	03:09:37:15	2:00:00	CU window/Jack in library

Figure 2-39 Edit decision list (long form) sample.

CONCLUSION

Television came first, with the invention of videotape coming over a decade later. Early videotape editing was the same as film editing, with editors cutting and splicing; but electronic editing changed all that, making it easier and cheaper to shoot and edit on videotape.

Television uses electronic pulses and interpretation of those pulses to generate a video signal. The television set is the receptor to which the converted signal is sent and transformed into a picture. A television picture is made up of two interlaced fields that scan the picture tube once every $\frac{1}{60}$ second. It takes two fields to make a frame and 30 frames to make 1 second of video by NTSC standard, the standard video color system used in the United States. Two other standards, PAL and SECAM, are used in other countries.

Video is magnetically recorded on one of five different sizes of videotape with one track of control track, one track to record video, and one or more to record the audio. SMPTE time code, a system to access specific shots according to their location on the tape designated by the hour, minute, second, and frame, may also be recorded on the videotape either in the vertical blanking interval or on an audio track. Time code can be drop frame, which reflects real time, and nondrop frame, which reflects TV time.

There are four basic camera framings, EWS, WS, MS, and CU, and six basic camera moves, static, pan, tilt, truck, dolly, and zoom. There are three basic angles from which to shoot, first person (POV), second person, or third person, and these can be shot from high, low, or straight forward to the subject. The way a shot is framed affects the way the program is edited and thus the way it is perceived by the audience.

Television aspect ratio is 3:4, with the safe title area being 80% of that and the safe action area, 90%.

A successful program is the product of a good plan. Among the lists that make for a good plan are the master checklist, schedule, budget worksheet, equipment list, crew list, cast list, location list, prop list, shot list, shot log, and the edit decision list. For ease of access, these lists and any other material needed to create the program should be kept filed in the program bible.

TO DO

1. Sketch the layout of tape showing the video track, the audio track, and the control track. Check your sketch against Figure 2–8.

2. Draw 20 squares, side by side. Let's say these 20 squares are the vertical blanking interval. What information would you store in these spaces? Check your answer against Figures 2–6 and 2–9.

3. Draw a box to the TV aspect ratio of 3:4. On this TV set, sketch in field

1 of a frame scan and then field 2 to demonstrate the interlacing between them. When you have both fields drawn in, what have you created? Check your sketch against Figure 2–4.

4.   How does SMPTE time code read on a tape? Check your answer against Figure 1–44.

5.   Where is longitudinal time code (LTC) recorded? Where is vertical interval time code (VITC) recorded? Check your answer against Figure 2–9 and see longitudinal time code in Chapter One.

6.   If one program is 22:37 long and a second is 22:39, but both will run 22:39, which was edited using drop frame time code? For the correct answer, see the end of the section on nondrop versus drop frame time code in this chapter.

7.   Make a shot list for the following scene. Include in your shots at least one each of the following: (1) CU; (2) MS; (3) WS; (4) a pan, tilt, zoom, or truck; (5) a POV shot; (6) a low- or high-angle shot; and (7) a head-on shot. Some of these can be combined. For example, a WS of the house could zoom to the front door or a POV shot could truck from the door to the car. Now you try it.

Scene: The living room of a house. The telephone rings. A man enters from a side door that leads to the kitchen, answers the phone, talks for a moment, and then hangs up the phone. He picks up his car keys from a nearby table, walks to the door, opens it, and exits. We see him as he hurries to his car and drives away with tires screeching.

A possible solution:

1. WS of living room to establish.
2. CU telephone as it rings.
3. MS (head on) of man entering from kitchen eating a doughnut.
4. ECU of man eating doughnut as he talks on phone.
5. TILT to follow CU on phone as he hangs up.
6. MS of man popping the last of the doughnut in his mouth as he walks over and picks up car keys.
7. POV TRUCK with man to door.
8. ZOOM IN to CU as he opens the door.
9. HOLD at door as man walks out and closes the door on the camera.
10. WS exterior of man leaving door and walking toward car.
11. Pull to high shot as he drives away.

3

Know Your Videotape

When it comes to selecting the best videotape for your recording or editing, there are a lot to choose from—some thirty different manufacturers, every one of them with five or six different grades, all with different specifications written in a language that only an engineer can decipher. Even so, it is important to work at finding the tape best suited to your system, not only for the original master recording but for the edit master as well.

You must consider the videotape's ability to withstand the stress of being shuttled back and forth in the play machine. In addition to fast forward and reverse transport, the tape is paused, stopped, rewound, loaded, and unloaded from the machine. All these things and more must be taken into account, particularly when choosing the videotape for the original material, since the quality of these shots will have a profound effect on the the final program.

The videotape used to record the edited master is subjected to some of the same physical demands, but to a somewhat lesser degree. Generally, it is not necessary to shuttle the edit master back and forth in the record machine as much, since you are generally working from one end of the tape to the other in sequence. However, the tape can be in pause for some length of time while the next shot is located or positioned to be edited in, and if the audio is mixed separately from the video, this procedure will subject the videotape to additional passes through the record machine, causing more wear and tear on the tape.

For all these reasons, you should use the very best videotape that is available for both the original material recording and the edit master. Go with known brands and ask the manufacturer for any testing information it might have regarding quality.

Generally, videotape that is designed to record more than 2 hours in the SP mode is *not* suitable for camera original recording or editing. The backing of these types of videotapes is thinner and is more prone to stretching. This becomes a real problem when you or your edit controller are trying to locate a specific frame for your edit-in and edit-out points.

Most manufacturers of videotape do a fair job of representing their products; however, they generally use terms such as *Professional Quality* or *New and Improved* or *Camera Grade* in their advertising or specifications. The things that you really need to know are not on the box or in any of the advertising. What you really need to know is whether the videotape is the best for your particular record and edit equipment.

There is really only one way to find out what is the best tape for your particular system and that is to test it. You can go through the trial and error scenario until you find just the right videotape for your system, but this is expensive and you are taking a risk, particularly if your production is a one-time event, such as a news conference or an interview. If the tape does not perform up to par, you are simply out of luck. The best way to test tape is to run it through the equipment and then evaluate its performance. It is not necessary to invest in a lot of test equipment or have an engineer evaluate it for you. Your own eyes are the most critical judge.

There are four basic things you need to know about the available videotape before you can select the right tape for your production. We will conduct four videotape tests to look for the following:

1. Color purity or smear
2. Edge sharpness and detail
3. Luminance capability
4. Amount of dropout

The first three have to do with bias and other technical and manufacturing factors. Bias makes sure that the videotape is magnetized over the range of the recorder head's magnetic characteristics. The fourth depends on the individual manufacturing process and/or the number of times the videotape has passed over the heads or been abused in storage, handling, or the recording process. The dropout test will be the last test. If "the best" has dropout problems, it will fail this test because there is no correction for this defect except to replace the videotape.

Before you start your videotape testing, set up your TV or monitor correctly so that what you are seeing is what you are recording. Because the videotape tests suggested here are subjective, your TV or monitor and your eyes are two of the three most important things that you will be using. The third is your own critical judgment.

Set up your TV or monitor properly by using the setup instructions supplied in Exhibits 3–1 through 3–4. Once your TV or monitor is set up so that you can evaluate your test results correctly, you are ready to begin testing your tape.

Exhibit 3–1   The Television Receiver or Monitor

There are three kinds of devices on which you can view your video production: (1) the standard TV set, (2) the monitor, or (3) the TV–monitor combination.

The **television set** is designed to receive television signals over the air or via cable using a **tuner** or channel selector. The broadcast station sends its signal through its transmitter over the air to the geographic area that it is licensed to cover by the FCC (Federal Communications Commission).

The TV station's broadcast signal is transmitted on a designated radio frequency (RF) band that is assigned a channel number that corresponds to a number on the tuner of the TV. All TV signals are received by an antenna that is connected to the TV set or by the cable hookup. When the set is turned on, the program appears on the picture tube and the sound is heard from the speaker.

Sometimes the cable company will change the channel number on their system from the one that the station is assigned, but the station is still broadcasting on its assigned radio frequency. All standard television sets are designed to receive only the **RF (radio frequency)** signal.

When you wire a VCR to a TV set through its antenna marked VHF output or UHF output, and you tune the TV to channel 3 or 4, the videotape from that VCR is played through the RF circuits in the VCR and the TV.

The **video monitor** does not have a tuner or channel selector. It is designed just to receive a video signal either from a VCR or a camera. A monitor may not have an audio amplifier or speaker. If it does not, it will not produce sound on its own but will need a separate amplifier and speakers.

The **television receiver–monitor** is a combination of a TV set and a monitor in a single cabinet with the video and RF electronics separated. The audio electronics work with both the video and the RF so that you can hear sound from the speaker whether you are viewing a broadcast signal (RF) or video from a VCR wired through either the monitor video/audio inputs or TV part of the set.

The Difference between Them

The difference between **RF** (the TV) and **video** (the monitor) can be seen in the clarity and sharpness of the picture. When a videotape is viewed on a monitor and is not processed through the RF electronics, the picture will be clearer and sharper than it will be on a TV set. The reason for the difference in the look of the picture is that RF is composed of several electronic signals that are carried with other signals over the air, while video only uses those necessary to see the picture.

In RF transmission, a TV station broadcasts its signal by combining the picture and audio information. These signals and others are combined in a device called a **modulator.** When the signal is received by your TV set, the video is processed through a **demodulator** that routes the picture information to the picture tube, the sync information to the vertical and horizontal circuits that hold the picture steady, and the sound to the audio amplifier, which then goes to the speaker(s).

The advantage of sending an RF signal is that we can send several separate TV signals through the air and they can be received over a single pair of wires, your antenna lead in or cable. The disadvantage is that you sacrifice picture quality and detail with the modulate, demodulate processing and the combining of all the necessary electronic information from several stations to send the picture and sound over the air or through the cable.

If, on the other hand, you have a monitor to view your production, you will see more of what you actually recorded with your camera than if you view your tape through a TV set.

Some of the ''super'' video formats, such as Super VHS (S-VHS), Extended Definition (ED), Hi-8, and Super Beta, separate the picture electronics information into parts called **chrominance** and **luminance** to give you more detail with increased resolution on a monitor that is somewhat the same as a professional model. On this type of monitor and VCR, the luminance is labeled **Y** and the chroma is labeled **C**. This is called component video.

Once you have set up your TV or monitor when you are testing videotape, shooting, or viewing your production, *don't change it*. It will tell you the truth, if the pictures are good or if they are bad and if the videotape is the best for your recorder. If you start changing things to make the picture better, the test will not be accurate.

Exhibit 3–2 How to Set up a Standard Television Set

To set up your TV, turn it on and let it warm up for about 5 minutes even if it is the instant-on type. While the set is warming up, turn off all the automatic controls that you can, including auto color adjustment and AFT (automatic fine tuning) or AFC (automatic frequency control). If you make your adjustments while these are on, they will override the manual adjustments to whatever the last settings were or will not change them at all. When you have finished the adjustments, you will turn them on again and the set will lock into the new settings.

If your set has a device that compensates for the amount of light in the room by adjusting the brightness of the picture, turn the room light on to your own standard for viewing. Then each time you look at your own production, set the room lights to this same level.

To make the setup as accurate as possible, choose a recorded program that meets the highest standards of video production technically. Or you can tune in the station broadcasting the best quality picture in your area. Be very critical when you select this station. If you have cable, there are differences in the quality of the signals from the stations. Select a program that is live, such as the live portion of a newscast, if you can, or select a soap or a prime-time program because they generally have the best production qualities. Adjust the fine tuning until you have the least grainy picture with the best definition. See Figure 3–1.

Now turn off the color completely. You want a black and white picture to set the brightness and contrast. When we say black, the black we mean is the darkest shade of gray that your set is capable of producing, and that is the shade that you see when the set is turned off. Many of you will think that you can see a darker black than when the picture tube is off, but, in reality, you can't. Remember that TV is small screen and that the lighter shades in the picture add to the apparent density of the shadows.

Turn the brightness down to where the screen is its darkest; then slowly turn it up until the darkest area of the picture is about the same shade of gray as the tube when the set is turned off. See Figure 3–2.

Look at the darkest area to see if there is any detail. If there is, then stop turning the brightness up. If you cannot see any detail in the darkest area, then turn the brightness up some more until you can. By detail, we mean in the part of the scene that is background, such as a chair in the corner

Figure 3-1 Adjust the fine-tuning until you have the least grainy picture with the best definition, as shown here.

Figure 3-2 When the TV is adjusted so that the darkest area of the picture is about the same shade of gray as the tube is when the set is turned off, the picture will look like this.

of the room. Can you see that it is a chair or is it just an object you can't identify? See Figure 3-3.

Next, turn the contrast all the way down until the whole scene looks dark gray. See Figure 3-4. Slowly turn the contrast up until you see the ob-

Figure 3-3 Turn up the brightness until you can see detail in the darkest area.

Figure 3-4   With the contrast turned all the way down, the picture will look gray or washed out.

jects clearly defined and there is a broad range of grays from the darkest (shadows) to the lightest, generally the main subject in the scene, with the lightest being comparatively white. See Figure 3–5.

Now turn the color on and up until the saturation looks normal. When

Figure 3-5   With the contrast fine-tuned, the objects in the picture are well defined with a broad range of grays from the darkest to the lightest.

you go past normal, the colors will appear to bleed over to other objects in the picture. Do not be concerned if the colors are not the right ones at this point; that is the next adjustment. If there is not enough color, the picture will look faded or washed out.

The next adjustment is the tint or hue of the picture. This control turns the picture from red to violet at one extreme to green at the other. We all know what color people are and that is your standard. Turn the control until a person on the screen has normal flesh tones.

Exhibit 3–3 How to Set up a Monitor

The most difficult part of setting up a monitor is finding a good quality prerecorded videotape to use as a standard. An alternative to videotape is to use the tuner in your VCR to receive a broadcast signal. To do this, hook up your VCR to the antenna or cable lead in, and then the video OUT on the VCR to the video IN on the monitor.

The set up procedure is the same as for a standard TV.

Exhibit 3–4 How to Set up a TV Receiver–Monitor Combination

Again, use the same procedure as outlined for the TV set if you choose to use a broadcast signal. If you choose to use a VCR, there is sometimes an overall brightness control on the set. It is there so that you can match the brightness of the VCR output to the RF or broadcast signal without having to adjust the set everytime you play a videotape. If there is a difference between the overall brightness of the broadcast signal and the VCR, adjust it before you make any other adjustments by using this control.

EVALUATING VIDEOTAPE

First, you will be performing three tests to determine the best tape for the original material. To do this you need a form to score the four or five brands and grades of videotape that you have selected. If you like, use the form supplied in the appendix and illustrated in Figure 3–6.

To be fair, it is best to conduct your test "blind" so that you are not influenced by the price, the manufacturer's name, or past experience. To do a blind test, have someone else take the videotapes out of their boxes and mark the boxes and the videotapes with an identifying number or letter; that is, mark one tape and box with an X and another tape and box with Y and so on. Once this is done, put the boxes away. You will use them to identify the tape after the test is complete.

When you select the brands and grades of videotape for your camera original recording, include one brand that you have used before and/or one that has been recommended by others. Also choose those with shorter recording times, if you have the option, that is, 20- or 30-minute cassettes, but never over 60 minutes. This will help to eliminate some of the tape stretch problem created by repeated shuttling in editing. In professional TV production, most field recorders will only take a 20-minute cassette and studio recorders are set up for 60 minutes.

Devise your own scoring system for the test, such as a 1 to 10 score for each videotape. Let's say the poorest score will be one, and the best, ten.

VIDEOTAPE EVALUATION TEST

For:

Master Material _____   Edit Master ID _____

Tape ID	Color Purity	Edge Sharpness	Luminance Capability	Total

Figure 3–6   This is the videotape evaluation test form you will need to complete the exercises in this chapter.

Test 1: Color Purity or Smear

For this test, you will need the following:

1. Your camera and VCR/VTR or camcorder and tripod
2. TV or monitor
3. Video cable to hook up the video out to the video in of the monitor *or* RF connector for your TV antenna terminal and the RF out of your VCR/VTR or camcorder
4. Four or five different brands of camera– or pro–grade blank videotapes for testing
5. Color video test chart (color bar chart) such as those made by Vertex Video Systems, Paso Robles, California
6. Two lighting instruments balanced for 3200 Kelvins (K) or your camcorder light mounted on the camera

This test should be in a room where you can close off all the natural daylight and artificial lights. You will be making judgments about color, so you will want to use only one color temperature of light for your test recording.

The color bar chart used in this test will have, from left to right, gray, yellow, cyan, green, magenta, red, and blue bars. If you do not have a color bar chart, make a workable facsimile with colored construction paper. If you do this, make sure that the edges are flush and butt up to each other so that they do not overlap and cause shadows on the adjacent color.

Set up the color bar chart on an easel or table adjacent to your TV or monitor. Set the camera on your tripod and make the necessary connections to your monitor or TV so that you can view the color bar chart on the screen.

Set the lighting instruments at about a 45 degree angle to each side and at an equal distance from each side of the bar chart. When you have completed your setup, it should look something like Figure 3–7.

Next, turn on the lights and frame the color bar chart so that it completely fills the viewfinder. Make sure that the image is perfectly square as seen in the viewfinder. This will confirm that each side and the top and bottom of the chart are all exactly the same distance from the camera lens.

Next, check to be sure that your lighting is flat and that there is no flare or hot spot on the bar chart or on your TV or monitor viewing screen. White balance your camera for artificial light.

Now compare the image on the screen with the real color chart and ask yourself if the colors match or if they are reasonably close. They will probably not match exactly, but you will want the two images to appear as nearly the same as possible for comparison. If the colors are way off between the chart and the screen image, check your lighting, the white balance on the camera, and the TV or monitor setup.

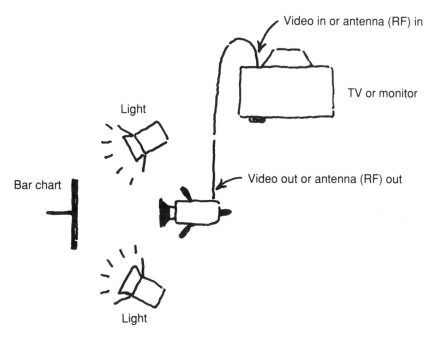

Video in or antenna (RF) in

TV or monitor

Light

Bar chart

Video out or antenna (RF) out

Light

Figure 3-7 At this point, your setup should look like this.

Next look at the edges of each color bar where they butt up against the adjacent color. What you are looking for is color bleed or the appearance of one color overlapping the edge of the one next to it. Depending on your camera and TV or monitor, you will note a slight bit of overlapping of the colors, generally on just one side.

Once the color is adjusted, the image will be your standard for judging the videotape's ability to record the colors with clarity and without smear. What you are seeing is the best image that your camera can "see" and your TV or monitor can display. The videotape recording will not improve the quality of this picture. In fact, the recorded image will probably not be as good as the real thing. So, what you see is what you are going to get. *The basis for your judgment will be how close the recorded picture is to the picture that you see through the camera and on the screen.*

The next step is to record about 3 or 4 minutes of the color bar image on each of your selected videotapes. After you have completed this step, it is time to make your evaluation by comparing each videotape's recording, judging the ability of the test videotapes to record the actual colors cleanly without a smeared or soft look. You are also looking for more overlap from bar to bar against your standard. See Figures 3-8 and 3-9.

Look at the videotapes one by one. Which has recorded the color bars best, with the best delineation between bars? This tape gets your highest score. The

Figure 3-8 If your color bars are overlapping, the colors will bleed into each other at the edges. In this black and white photo, you see this as a definitive line between colors. Compare this illustration to Figure 3-9.

scores of the other tapes are based on the score you give this one. If your highest score is an 8, with 10 a perfect score, then your second best tape will get a score below 8.

You must be very critical when making this judgment to note the smallest differences, since as the pictures go down in videotape generations, the smallest deficiency will become more apparent and objectionable. After you have evaluated and assigned a score to each videotape, your evaluation form will look something like Figure 3-10.

Test 2: Edge Sharpness and Detail

The setup for this test will be the same as for the color purity and smear test. The color bar chart will be replaced with a page from the newspaper. Select a page

Figure 3-9 Color bars butted up to each other properly will have little color bleed except that generated by the camera or the monitor. Compare this illustration to Figure 3-8.

VIDEOTAPE EVALUATION TEST

For:

Master Material ×   Edit Master ID

Tape ID	Color Purity	Edge Sharpness	Luminance Capability	Total
1	7			
2	9			
3	6			

Figure 3-10 After you have evaluated and assigned a score for purity to each videotape, your videotape evaluation test might look like this.

with a variety of type sizes printed on it. You will record this image immediately following the color bar image.

For this text, you will need the following:

1. Your camera and VCR/VTR or camcorder and tripod
2. TV or monitor
3. Video cable to hook up the video out to the video in of the monitor *or* RF connector for your TV antenna terminal and the RF out of your VCR/VTR or camcorder
4. Your test videotapes with color bars already recorded on them
5. One page from a newspaper with a variety of type sizes
6. Two lighting instruments balanced for 3200 K or your camcorder light mounted on the camera

First, tape the page from the newspaper to a wall to make it as flat as you can. Place the lights as you did in test one and check to see that you do not have a hot spot reflected from the wall back toward the camera. When you are finished with your setup, it should look like Figure 3–11.

Make sure that the camera lens is square with the newspaper. Frame a tight shot of the whole page and check the focus to see that everything is sharp. Set the zoom lens to wide-angle and do not change it.

Next, mark the position of the tripod with a piece of tape on the floor. Now look at the image on your monitor. This image is your standard, as it was for the previous color purity test.

Record about 1 minute of video on each of your test videotapes behind the color bar image. Then move the tripod back about 12 inches and mark this position on the floor with tape. Again record about 1 minute of video behind the first newspaper recording on each of your test videotapes. Continue to move your

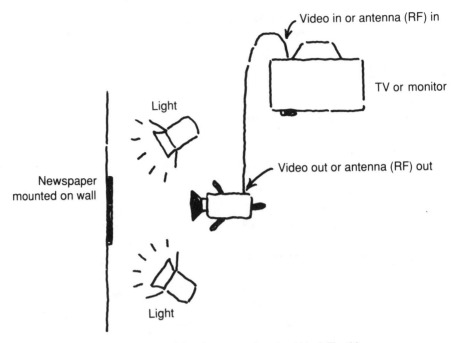

Video in or antenna (RF) in

TV or monitor

Light

Video out or antenna (RF) out

Newspaper
mounted on wall

Light

Figure 3-11   At this point, your setup should look like this.

tripod back until you can no longer read the largest type on your monitor, recording on all tapes after each camera movement.

When judging the recorded image, look for the one that provides the best definition of the edges of the letters and the greatest clarity of the type in the various type sizes. Give this videotape the highest score. See Figure 3-12.

Also judge the clarity from the different distances as the camera was moved back. The best performer gets the best score. See Figure 3-13, 3-14, and 3-15. After completing these tests, your evaluation sheet might look like Figure 3-16.

Test 3: Luminance Capability

This test will show you the range of luminance and dark colors the videotape is capable of recording.

For this test, you will need the following:

1. Your camera and VCR/VTR or camcorder and tripod
2. TV or monitor
3. Video cable to hook up the video out to the video in of the monitor *or* RF connector for your TV antenna terminal and the RF out of your VCR/VTR or camcorder

Figure 3-12 When judging the recorded image of the newspaper, give the highest score to the one with the best definition of the edges of the letters and the clarity of type. (Newspaper courtesy of the *Los Angeles Times*.)

Figure 3-13 Judge the clarity of type as the camera is moved back. (Newspaper courtesy of the *Los Angeles Times*.)

Figure 3-14   Continue to judge the clarity of type as the camera moves even farther back. (Newspaper courtesy of the *Los Angeles Times*.)

Figure 3-15   Judge the clarity of type when the camera is as far away as it will be. (Newspaper courtesy of the *Los Angeles Times*.)

VIDEOTAPE EVALUATION TEST

For:

Master Material ×　　　　　　　　　Edit Master ID

Tape ID	Color Purity	Edge Sharpness	Luminance Capability	Total
1	7	8		
2	9	7		
3	6	9		

Figure 3-16 After you have evaluated and assigned a score for sharpness to each videotape, your videotape evaluation test might look like this.

4. Your test videotapes with color bars and the newspaper recordings already on them
5. A gray scale test chart similar to ACCU-CHART® 11 Step Gray Scale available from Vertex video systems, Paso Robles, California
6. Five sheets of construction paper: black, dark gray, dark blue, dark brown, and medium brown
7. Two lighting instruments balanced for 3200 K or your camcorder light mounted on the camera

First, set up your camera or camcorder, monitor, and lights as you did for the previous test. Then mount the gray scale test chart on the wall or on an easel. The lighting for this test is critical. It should be as flat as you can possibly make it, which means that it must be even overall with no hot spots or shadows. White balance your camera and frame a tight shot of the chart. The suggested gray scale chart has 11 steps from 90% reflectance to 2% reflectance. See Figure 3-17.

The next step is to observe the image on your monitor or TV. The picture that you want to see is each gray color step, from the darkest black to the lightest white. You might have to adjust the iris on the camera if you cannot see the full range. You will be pushing your camera and iris to their limits because the suggested chart is about 5.5 F-stops, the technical limits of the video system. Normally, you would not videotape a shot with this range of light reflectance.

If the darkest bars are not distinct, open the iris a little. If, on the other hand, the lightest bars are blooming, close the iris slightly. Keep adjusting the iris until each bar on the chart is discernible. After this step is completed, the image that you see on your monitor or TV is your standard, as it was on the previous test. Now that you are set up, record about 1 minute of the gray scale on

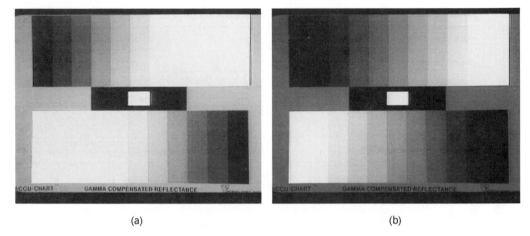

(a)   (b)

Figure 3-17 To test the range of luminance and dark colors that the videotape is capable of recording, frame a tight shot of a gray scale chart. In photo (a) the camera iris should be closed down until you can see the delineation between the shades (reflectance) of gray, as shown in photo (b).

each of your test videotapes *after* the image that you recorded for the sharpness test.

Once you have completed that, set up to record the construction paper by laying out each different color of paper, starting with the darkest and progressing to the lightest, so that you have about $1\frac{1}{2}$ inches of each color showing. You can tape the sheets together on the backside to keep them in place if need be. See Figure 3-18.

Replace the gray scale chart with the assembled sheets of construction paper. Check your lighting and white balance the camera. Check your iris setting

Figure 3-18 Your construction paper layout should look like this, progressing from the darkest to the lightest color.

to see that it is in the normal setting. Look at the image in your monitor. Again, this is your standard for evaluation in this test. See Figure 3–19. Record about 1 minute of the different colors of construction paper after the gray scale recording.

When you observe the results of this test, look for the maximum delineation of the gray scale bars and the maximum color definition of the construction paper before the darkest color is no longer distinguishable from the black sheet of paper. See Figures 3–20 and 3–21. After you have completed this test and totaled the test results, your evaluation sheet might look like Figure 3–22.

LOOKING AT TEST RESULTS

You have now completed your evaluation for the best videotape for recording your camera original. As you can see in our hypothetical test shown in Figure 3–22, videotape 1 was the best overall, although not a clear winner. The closeness of the results could be a function of the way we remembered our standards and the scores that we gave each videotape. You must remember that this test is strictly subjective, and the things you see on one day will not necessarily be the same on another. This statement is true for most of the video production and post-production process.

This selection of the best videotape for recording the camera original is only one-half of the requirements for videotape in the total process of production and editing. The other half is to test for the best videotape, according to your test, for your edit master. It could end up being the same as the videotape that you use for the camera original or it might be different. Following is how to test for the best videotape to use for the edit master.

Figure 3–19 Record about 1 minute of the construction paper setup once you have it focused properly.

Figure 3-20   If the construction paper setup looks like this, adjust the camera iris until you see maximum delineation between colors.

TESTING THE EDIT MASTER

To find the best videotape for your edit master, we use basically the same process with one change. Instead of using your camera or camcorder as the video source, use the tape you decided was the best based on the previous test. Dub this video over to a second selection of brands and grades of videotape for your edit master. The videotapes you use for this test can be new tapes or can include some of the ones that you used in tests 1 through 3. Just because a tape scored low on the previous test does not necessarily mean it will score low on this one. Certainly you should include a new tape from the manufacturer of the winner from the previous test. It only makes sense to try the best.

For this test, you will need the following:

1. Your player for the selected videotape
2. Record machine

Figure 3-21   This shows maximum delineation between colors of construction paper.

VIDEOTAPE EVALUATION TEST

For:

Master Material × _____   Edit Master ID _____

Tape ID	Color Purity	Edge Sharpness	Luminance Capability	Total
1	7	8	9	24
2	9	7	6	22
3	6	9	8	23

Figure 3-22 After you have evaluated and assigned a score for luminance and totaled all the scores, your videotape evaluation test might look like this.

3. TV or monitor

4. The winning videotape from the previous tests (tape 1 in our example)

5. New blank videotapes or the ones you used before

First connect the video out of the player to the video in of the record machine. Connect the video or RF in of the TV or monitor to the video or RF out of the player. See Figure 3-23.

Now turn on the edit switch of the *player* if it has this feature. Next turn the system on and insert the best camera original videotape in the player (tape 1

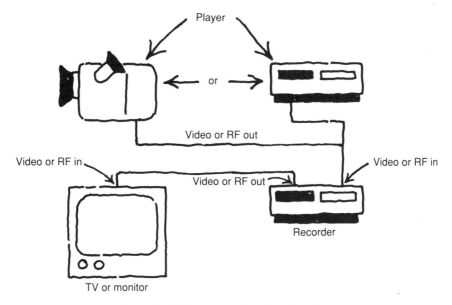

Figure 3-23 Your setup should look like this.

from the previous test), press play, and observe the image that you see on your TV or monitor. The picture you see will become your standard for the edit master videotape selection. This will be the best picture that your player, TV or monitor, and master material will be able to play back for the edit master to record.

Before you begin the test, give all the tapes to be tested a letter designation as you did before: A, B, C, and so on. Insert your selected videotapes one at a time and make a dub of the winning tape from the previous test onto each of them.

Observe the results, noting on a new videotape evaluation test form a score for color purity and smear, edge sharpness and detail, and luminance as in the previous test. Add up the scores to determine the winner. The comparison of the four or five dubs will show you the best videotape to use for your edit master. See Figure 3–24.

Test 4: Amount of Dropout

Dropout exists where there is not sufficient oxide coating on the tape to record the video signal. Dropout appears as white splotches in the picture when the videotape is viewed as seen in Figure 3–25. All videotape, regardless of its quality, has some dropout.

Generally, videotape with "pro" or "camera" designations has an extremely smooth magnetic coating, and the oxide particles are densely packed on a more substantial backing than for normal-use designations. Because these grades have more oxide, there are less dropout problems in the manufacturing process.

Other than defects in the manufacturing process, the most prevalent cause of dropout is leaving the tape parked in pause for an extended length of time or parking it in pause at the same spot over and over. This causes the video head to

VIDEOTAPE EVALUATION TEST

For:

Master Material ×_____   Edit Master ID_____

Tape ID	Color Purity	Edge Sharpness	Luminance Capability	Total
A	7	8	7	22
B	9	9	7	25
C	6	9	6	21

Figure 3-24 Your videotape evaluation test to determine the best videotape to use for the edit master might look like this when it is complete.

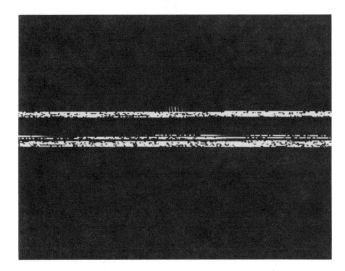

Figure 3-25 Dropout appears as white splotches on the tape. This photo shows dropout against video black.

eventually wear down the amount of recording material at that particular spot, causing dropout.

Here is how to do a simple test for dropout. For this test, you will need the following:

1. Your camcorder or recording VCR/VTR
2. A video black source. A SEG with a fade-to-black feature or your capped camera or camcorder will work just fine for this test.
3. A video monitor or TV
4. Both winning videotapes from the master material tests (tape 1) and the edit master test (tape B)

Make the necessary connections to hook your video black source to the record machine as shown in Figure 3-26. Load your rewound videotape into your recorder and record black on the videotape over *all* recorded material, recording black over everything recorded in the other tests.

When the tape is recorded black, play it back and watch the black screen on your monitor for the number of white splotches that appear and their duration. Keep in mind that the video recording is 30 separate pictures each second. Generally, minimal dropout will not affect the look of the final production, unless it is followed by more. If it lasts beyond a few seconds or there is a lot in one part of the tape, you may not want to use this manufacturers's tape as your edit master.

When doing your evaluation, pay particular attention to the head end of the videotape. This is where you recorded your camera original and edit master test, which means that the videotape has been recorded on at least once. And if there is a manufacturing problem, this is where it will more than likely show up.

Black lens cap in place

External mike terminal with
adapter jack insterted

Figure 3-26 If you are using a camera
as your source for video black, your
setup should look like this.

By doing this test on several different videotapes and on one with known
dropout problems, you will be able to make a quantitative judgment on whether
a particular videotape has too much dropout to trust with your edit master.

AVOIDING TAPE PROBLEMS

Videotape problems caused by mishandling can be avoided by following a few
basic tips:

1. Remove cassette tabs or buttons to avoid accidental erasure. All cassettes have these tabs or buttons, which break off or pop out. Reel to reel tapes
do not. See Figure 3-27.

Figure 3-27 Three-quarter-inch cassettes (pictured right) have buttons
that can be removed to avoid acciden-
tal erasure, $\frac{1}{2}$-inch cassettes have tabs
that can be broken off, and 8mm has a
tab that can be slid inside the cassette.

2.   Label cassettes or reels *before* you start to roll tape. Since you cannot see the recorded pictures, this will help stop accidental erasure by recording over material.

3.   Store your cassettes or reels in a cool, dry place. Heat and moisture will damage tape.

4.   Check for dropout *before* you begin to record. Overused tapes may have lost some of the oxide coating and will thus leave holes in the video.

5.   Don't leave tapes in pause for a long period of time. Part of the magnetic field may be destroyed and, with it, video and audio.

6.   Store tape in boxes to minimize dust contamination. One particle of dust can cause a glitch in the record process.

7.   Fast forward to the end and then rewind to the beginning on all tapes before recording to establish uniform tension.

These tips will not prevent all problems, but they will at least get you started in the right direction.

CONCLUSION

A videotape evaluation is not a technical test of all the electronic signals that are recorded on the videotape. The test is very subjective, as is the whole process of producing a TV program. However, the test makes it possible for you to trust the videotape that you are using and be assured that you are recording the original material and the edit master on the best videotape available for your equipment.

Once you have found the right tape, treat it right. Store and handle it properly. Good tape habits both before, during, and after the taping will produce top-quality original material and lasting edit masters.

TO DO

If you did not do the videotape test, try it. You might test the tape you usually purchase with another brand. You may discover that something better is available.

4

How TV Is Edited

Post production is the final phase in the creation of a television program. It includes all activities that follow the completion of taping up to the finished program, including, but not limited to, the selection of shots and deciding on transitions between those shots; the selection of music, audio, and sound effects; the selection of graphics or other visual elements; and, finally, the execution of these decisions in an edit session. All these activities are preparation for the edit or execution of the edit and, because of this, post production and editing are often thought of as being synonymous. Thus, the edit may be called post.

The ultimate goal of the post-production process is to transform hours of raw footage into a concise, effective format that is designed specifically to most effectively tell the story or deliver the message to the audience. All programs require some form of editing, even if it is only to add a title or music. Game shows, for example, require very little editing since they are taped using multiple cameras switched live by the director. However, they may require some pull-up to conform to time constraints, and titles, credits, and prizes may be added in post. On the other hand, most story-based programs require extensive editing and thus become a product of the editor's expertise, as well as the producer's, the director's, and the writer's.

Editors may have an impact on what shots are used and what transitions hold them together; how long a shot lasts and what follows; what audio is kept from the original footage and what is added; and, most critically, the overall pacing of the program. In short, an editor can have a tremendous effect on the success of a program based on the way he or she puts it together in post.

There are those who believe that a bad program can be saved in post, and it is not uncommon to hear someone say, "Don't worry. We'll fix it in post." The philosophy has *some* merit. A good editor can make a program better by not editing in poor shots, bad audio, ineffective angles, and just all around unusable video, but even a good editor cannot save a program that has no redeeming elements. There is a certain amount of magic in the edit, and a good editor that understands the mechanics (meaning the equipment) as well as the techniques (meaning the step by step process) can do much to make a program both entertaining and effective.

ASSEMBLE VERSUS INSERT EDITS

There are two basic editing methods: the assemble edit, shown in Figure 4–1, and the insert edit, shown in Figure 4–2. Any edit setup can do an assemble edit, which is simply the sequential recording of the edit master from the front to the back. Everytime you do an assemble edit you record video, audio, and control track.

Insert editing, on the other hand, is not a feature of all edit systems. An insert edit is, as its name indicates, the ability to edit video and/or audio in between existing shots without disturbing those existing shots *and* without disturbing the recorded control track. In other words, you erase only that portion of the video and/or audio that you want to, inserting the new video and/or audio in its place.

An insert edit can be:

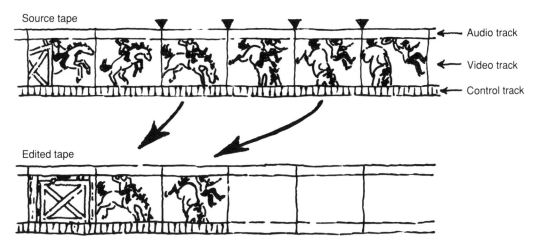

Figure 4-1 An assemble edit records the edit master sequentially, from the beginning to the end of the program. An assemble edit records new video, audio, and control track on every edit.

Figure 4-2 An insert edit records video and/or audio *between* existing video and/or audio without disturbing the existing video and/or audio surrounding it *and* without disturbing the recorded control track.

1. V or video only
2. V + A1 or video plus audio track 1
3. V + A2 or video plus audio track 2
4. V + A1/A2 or video plus both audio tracks
5. A1 or audio track 1 only
6. A2 or audio track 2 only; or A1/A2 or both audio tracks, but no video

The audio tracks may be designated as left (L) and right (R) instead of 1 and 2.

With an insert edit, you can slip a piece of audio under existing video or you can record music behind the entire program. You can also change a shot that may not be working by telling your edit system the exact edit-in and edit-out points and insert a new one.

Which to Use

If you have the choice, there are definite advantages to using insert edits. First, an insert edit does not disturb the recorded control track, whereas an assemble edit does. An assemble edit records new control track with every edit, and this can cause glitches where the new and the old control track meet. An insert edit does not have this problem. If you are doing insert edits only, you can record control track from one end of the tape to the other by blacking the tape and never have to worry about it again. The insert edits will not disturb it. With continuous control track like this, the tape will play better and thus record better.

Split Edits

A split edit is when you give the video and the audio different edit-in or edit-out points so that one leads or trails the other. For example, if you are going from a wide shot to a closeup and you want the audio from the closeup to loop under the wide shot, you would split the edit by telling the audio to begin recording sooner than the video.

Split edits are generally available only on state of the art editing systems. However, you can do one in two passes on a basic system that has insert edit capabilities and control track editing. Using the example above, you would make your audio edit first, recording it up to the point where the closeup will come in. Then, on the second pass, you record both audio and video, matching the outgoing frame of the wide shot. A match frame is just that—making the edit in at the exact frame where you made the edit out so that no frames are lost in the edit. Make your frame match using time code or control track numbers. Don't try to do it visually or by ear unless there is just no other way. In this example, the frame match would be an audio only since the video is changing on the cut.

If this had been music instead of on-camera dialog, you could have laid all the music in the first edit and then laid only the video in the second edit. You never want to try to cut music together anyway. It is just too hard to match exactly.

PLANNING AN EDIT

The secret to a successful edit is a well-planned and well-documented shoot and a good editing plan. If you do not have these, you will spend your edit time running tape back and forth, undecided where to begin an edit, undecided about what shot to use, and inevitability editing in a shot only to discover later that there is another one that would have worked better.

It is critical to know where you are going. Visualize in your mind what your edited program will look like. Know what is recorded on your original material. Decide what will go first and what follows. In other words, have a plan.

Three planning elements are absolutely essential to successful editing:

1. Concept
2. Shot logs
3. Edit decision list (EDL)

The concept gives you direction, the shot logs tell you what you have on tape, and the edit decision list (EDL) takes you step by step through how you have decided to put it all together.

Concept

The post-production and edit plan of all TV productions starts with the concept. It really does not matter if you are producing a 3-hour prime-time special or a 30-second commercial. Post production and editing starts before you ever load the videotape or push the record button.

Today's TV production equipment has the capability to create some great effects, but the reality is that continuous effects can become boring quickly. It is ultimately the content of the program that will hold the viewer, and the content of a video production takes planning—not just an editing plan but planning from the very beginning. The basis for the plan begins with a script.

In broadcast TV productions, regardless of the subject, a script is an absolute necessity. If there is no script, there is no show or commercial or newscast or game show or station promotion or anything else. Why? The biggest reason is that without it no one would know exactly what the program is or what it is supposed to look like except the person who created it. The on-camera or voice-over people would not know what to say or do. The director would not be able to direct the actors or be in a position to tell the crew where to be and what they are supposed to do. And finally, and most important in broadcast TV, no one would invest the money to produce a program without a script because there would be no way to determine how much it was going to cost. That's just not good business.

Not all programs require a complete script, but all do require extensive thought and some basic planning. An interview show, for example, may be able to get by with just a scripted open and close, with an outline of questions to be asked inside the program. In this case, the program is essentially created live by the interviewer and interviewee. A game show is another good example of a partially scripted show. Game shows have a basic script, which includes the "must say" dialog and outlined portions that are unscriptable, for example, contestant responses (it cannot be predicted how a contestant will respond when asked to tell the audience something about him or herself).

A dramatic program, on the other hand, is totally scripted, with nothing said or done that is not included in the script. There are some exceptions to this hard and fast rule. The story can be changed slightly, a scene added or deleted, dialog added or deleted, and so on, while the program is being taped based on the input of the producer, the director, or the performers. Going in, however, the plan is to shoot the script as is.

A script then can be complete or a basic outline, but in either case it is a step by step process that leads to the completed program, executed so that it is a duplicate of what its creator had in mind in the beginning. There are two different script forms. The form shown in Figure 4–3 divides the page in half, putting the video and any sound effects or music on the left side and the spoken word (dialog) on the right side. This type of script is used for shorter productions, such as commercials. The form shown in Figure 4–4 was developed for longer programs. It breaks the program down into scenes, with video and sound effects

SCRIPT

STYLE 1

VIDEO OR EFX. ALL CAPS NOT
UNDERLINED. XXXXXXXXXXXXXXX
XXXXXXXXXXXXXXX SOUND/MUSIC/
AUDIO EFX: ALL CAPS. UNDERLINED.
XXXXXXXXXXXXXXXXX

STAR'S NAME. UNDERLINED.
Dialogue. Upper/lower case.

    Xxxxxxxxxxxxxxx. Xxxxxxxxxxx.
Xxxxxx. Xxxxxxxxxxxxxxxxxxxxxxx
    xxxxxxxxxxxxxxxxxxxxxxxxxxxxxx
    xxxxxxxxxxxxxxxxxxxx. Xxxxxxxxxx
    xxxxxxxxxxxxxxxxxxxxxxxxxxxxx.

Figure 4-3 The short script divides the page in half, putting video and any sound effects or music on the left side and the spoken word (dialog) on the right side. (James R. Caruso/Mavis E. Arthur, A BEGINNER'S GUIDE TO PRODUCING TV© 1990, p. 152. Reprinted by permission of Prentice Hall, Englewood Cliffs, New Jersey.)

and music noted in all capital letters, margin to margin, and dialog indented and in upper- and lowercase.

    An addition to the script may be a storyboard. See Figure 4–5. Storyboards are an artistic rendition of exactly what a scene should look like and may or may not include any dialog, music, or sound effects that go along with the scene. A storyboard can be basic, showing only the base scene, or it can be specific, with

STYLE 2

VIDEO/VIDEO EFX: ALL CAPS. NOT UNDERLINED.XXXXXXXXXXXXXXXXXXXX
XXXXXXXXXXXXXXXXXXXXXXXXXXXXXXXXXXXXXXXXXXXXXXXXXXXXXXXXX

SOUND/MUSIC/AUDIO EFX: ALL CAPS. UNDERLINED. XXXXXXXXXXXXXXXX
XXXXXXXXXXXXXXXXXXXXXXXXXXXXXXXXXXXXXXXXXXXXXXXXXXXXXXXXXX

              STAR'S NAME. UNDERLINED

           Dialogue. Upper/lower case. Xxxxxxxxxxxxxx

           xxxx xxxxxxxxxxx xx xxxxxxxx. Xxxxx xxx

Figure 4-4 The longer script breaks the program down into scenes, with the video and sound effects and music noted in all capital letters, margin to margin, and dialog indented and in upper- and lowercase. (James R. Caruso/Mavis E. Arthur, A BEGINNER'S GUIDE TO PRODUCING TV© 1990, p. 152. Reprinted by permission of Prentice Hall, Englewood Cliffs, New Jersey.)

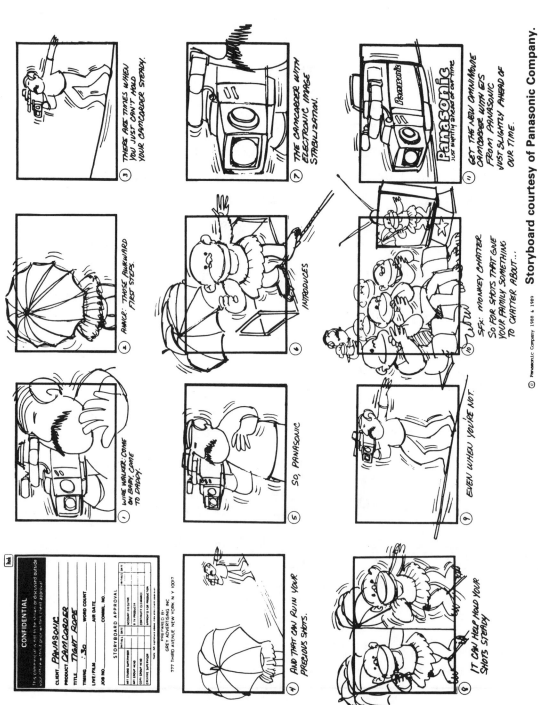

Figure 4-5 Storyboards are an artistic rendition of exactly what a scene should look like and may or may not include any dialog, music, or sound effects that go along with the scene. (Storyboard courtesy of Panasonic Company.)

a new panel for every camera angle or every major move by the actors. Story-boards are typically done for commercials but some film makers use them extensively as well. Alfred Hitchcock storyboarded every scene of every movie he ever made, and Steven Spielberg uses storyboards extensively in his movies.

Shot Logs

A shot log is a list of every shot on a piece of tape. It includes the shot number, counter or time code location, an indication of whether audio has been recorded, and a place to make notes about the shot, for example, good, NG (no good), too long, too short, or sloppy, and any other information that might be helpful when preparing the EDL. See Figure 4–6.

Comprehensive shot logs are absolutely essential to a successful edit. These logs are the only way of knowing what is recorded on the videotape, short of sitting down and looking at the material, a time-consuming process that no editor wants to do or should ever have to do.

A shot log is prepared by first giving every scene a number. This number is usually designated on the script and written on the shot log when a particular scene is taped. This facilitates preparing the EDL and putting the final program together. Number scenes consecutively, with cutaways coded to the main scene. For example, a wide shot of a cruise ship that will precede a cabin scene may be numbered 22 with the scene that follows it 23. See Figure 4–7. After you have recorded scene 23, you may elect to record a closeup of an actor, perhaps to get a better shot of a reaction. Number this shot 23-A. This way you immediately know that the closeup belongs with scene 23. It's that simple. As a matter of fact, the simpler the system is, the better. Do not try to complicate the process.

Producing a television program is like constructing a building. On the whole it seems an enormous task, but if you take it one step at a time, reducing it to its simplest form, the project becomes relatively easy.

Edit Decision List

Something as important as an edit decision list probably should have its own chapter, even its own book. However, it would be a very short chapter and one of the shortest books ever printed because, while its importance is significant, the process is simple.

An EDL is nothing more than a consecutive list of what will happen in the program, both visual and audio elements, including opening titles, graphics, sound, music, shots, and animation. The EDL is the program on paper and thus sometimes it is called a *paper edit.*

In its finished form, the EDL is usually just paper with numbers and vague descriptions written, and it may be understandable only to the person that compiled it. It looks simple, but it requires a lot of time and a lot of thinking to make it a good one. At its best, an EDL should allow you to go into the edit session

SHOT LOG

LOCATION: _____   Show: _____

Tape No.	Shot No.	Take No.	Time code/counter numbers		Audio	Notes
			In	Out		

Figure 4-6 Shot log form.

22   SHIP AT SEA - DAY

                                                              DISSOLVE TO:

23   INT. JAN'S CABIN - DAY

          Jonathon is crouched with his ear to the adjoining door to Heather's room. Jan is
          pacing behind him.

                                    PAM
                        I don't believe this.

Figure 4-7 A shooting script breaks the scenes down into shots. Note that the
one pictured includes a wide shot to establish location before dissolving to the
actual scene location, a cabin aboard ship.

totally ready, with all the basic decisions made and the location of everything
known. It should take care of the mechanics of the edit so you can concentrate on
the creative aspects of editing the program together. If you are new to the process, an easy way to compile an effective EDL is to begin with 4 by 6 inch cards.

Using the shot logs, all shots you are considering for the program are written onto 4 by 6 cards. Make a card for *every* shot you are considering for the
video. This does not include every shot you recorded, but only those you really
believe you might want to use as a part of the program. Once these cards are
made, you can easily arrange the shots based on how they develop the concept
of the program. It also becomes easy to check continuity and decide on transitions between shots with this card layout.

After you have the EDL cards in the edit order, go back through and number them consecutively. Then go through them once more, writing in the transitions from shot to shot, keeping in mind continuity, pacing, and timing.

EDL cards can be taken directly to the edit or they can be used to make a
paper EDL. A paper EDL includes the same information as the card EDL except
that it is listed in consecutive order on paper. Both have advantages. A paper
EDL, for example, cannot be changed easily. A card EDL can. Just move the
cards around. To the paper EDL's advantage, there is the question of whether it
might be easy to misplace cards. Probably the best solution is to make both a
card EDL and a paper EDL and then you have the flexibility of using either.

A paper EDL can be prepared on a form like the one shown in Figure 4-8
and a card EDL prepared on a form like the one shown in Figure 4-9. Both include the following information:

1. **Scene:** Write here the shot number as noted on the script. This will key this
   shot back to its place within the script.
2. **Tape number:** Every videotape you will be pulling video and/or audio from
   should have its own number for easy reference. Tapes should be numbered

either consecutively or by program number or date shot or scene number or something that makes sense in terms of the program. *Note:* Time Code is often keyed to a tape number. For example, if the tape is number 6, the time code will designate the hour as 06, that is, 06:00:00:00. This way you immediately know what tape you have up by just looking at the hour on the time code. See Figure 4–10.

3. **Shot number:** This is the number of the shot as designated on the shot log.

4. **Edit type:** There are three kinds of edits: (a) audio only; (b) video only; (c) audio and video. If your equipment has the ability to access two audio channels (either left and right or 1 and 2), you will need to designate this also, for example, A1 for audio channel 1 and A2 for audio channel 2. If you write only A, then the audio is recording on both channels. What kind of edit are you going to make? Write A or A1 or A2 for audio only, V for video only, and A/V or A1/V or A2/V for audio and video edit simultaneously from the same source.

5. **Transition:** How are you going to get from shot to shot? Will you do it with a cut, dissolve, fade, wipe, special effect, or what? Be sure you know what your equipment can do so that you do not plan a transition you cannot possibly execute. Also, if you are doing a dissolve, fade up or down, or special effect, indicate the transition time. Write in this column one of the following: Cut, Dissolve/:30 (meaning a 30 frame dissolve); Fade up/:10 (meaning a 10-frame fade up); or EFX (meaning an effect). If you use an EFX, describe the effect in the description column.

6. **Time code numbers or counter numbers:** The exact location of the shot you want to use (in point and out point) or at least as close as you can get, given the equipment with which you are working.

7. **Duration:** How long is the edit? This is helpful for timing purposes. For example, write 5:00, meaning 5 seconds, or 00:10, meaning 10 frames, or whatever.

8. **Description:** Make any notes on the shot, including framing, for example, CU, WS, MCU, NARRATION, EFFECT, 3 SHOT; shooting comments; and so on.

An edit decision list prepares you to edit. Every sound and visual element of your production becomes a part of the EDL—music, graphics, video, dialog, whatever. If it is going into the program, it is on the EDL. For example, if you want to add a musical background, this is noted as an edit on the EDL at the place where the music is faded in. The same holds true for a graphic like a title that is computer generated or comes by way of a SEG (special-effects generator). This is an edit just like a picture from the recorded videotape and belongs on the EDL.

It is accurate to say that the EDL is a consecutive list of what happens in the program, whether the EDL is a series of cards or a single form. It is also true that

EDIT DECISION LIST

Scene	Tape No.	Shot No.	Edit Type	Transition	Time code/counter numbers (tape no. is designated hour)		Duration	Description
					In	Out		

Figure 4-8 Edit decision list (EDL) long (paper) form.

127

EDL FOR SHOW

Scene no.	Tape no.	Shot no.	Edit type	Transition
_____	_____	_____	_____	_____

Time code/counter numbers

In	Out	Duration
_____	_____	_____

Description

Figure 4-9 Edit decision list (EDL) single-entry, card form.

some things that happen may continue through the next edit. See Figure 4–11 for an example of what you might find on an edit list.

Obviously, we are not going to play music for 2 minutes over a black picture. That would not be very good TV. What we are going to do is lay the music in for 2 minutes and edit different video over it. It may take several edits to provide pictures for the music, but you will know this by the EDL. See Figure 4–12.

When the music ends, some other audio will take its place unless we want pictures over no sound at all. The edit decision list is of absolute importance; without it, you edit blindly. Do not underestimate its importance. Such folly will cost you time and money.

CONTINUITY, PACING, AND TIMING

In Chapter Two we stressed the importance of planning camera framing, angles, and moves with the edit session in mind. The significance of this becomes obvi-

Figure 4-10 Number all tapes. The tape number may correspond to the hour time code to make the shot logs and the edit decision list easier to use.

ous as you prepare an EDL and discover that shot 4 pans to the left at its end and shot 5 pans to the right as it begins, and they are supposed to be cut together. If that is not enough of a problem, shot 6 tilts up at the beginning. Cutting these three shots together and making them work for the viewer is going to a real challenge, if not an impossibility. You cannot pan left, then pan right, and then tilt up unless you are on a roller coaster, because that is what it will look like to the viewer. The same holds true for action on the screen. For example, it is not going to work for the viewer if a person is walking toward a house and in the next shot the person is inside the house. How did the person get inside? Or if a person is wearing a hat in a scene, but when you cut to another angle, the hat is gone. Where did the hat go? Things like these can ruin a program for the viewer.

When the shots are already on tape, it becomes the editor's problem. If you handle it when you plan the shots, it will not be a problem at all, but a conscious decision to shoot it the right way to begin with. Ultimately, whatever is on tape is all the editor has to work with to put the program together. It is the editor who must deal with whether the shots are there to make a good program.

The editor is the last link in the chain. It is his or her job to make sure the edits work, not to just cut the show together without considering how shots fit together. Ultimately, it is up to the editor to exclude shots that do not work with something that will. Major considerations for the editor are continuity, pacing, and timing.

Continuity

Never cut two shots together that, because of their content or framing, distract from the program. In other words, create smooth transitions from one shot to the next. Shots should flow logically not only as far as the storyline goes, but also as far as the visual goes. For example:

EDIT DECISION LIST

Scene	Tape No.	Shot No.	Edit Type	Transition	Time code/counter numbers		Duration	Description
					In	Out		
					(tape no. is designated hour)			
	6		AI	FADE UP	00:21:23:25	00:21:25:22	01:27	MUSIC

Figure 4-11 This is an example of what you might find on an edit decision list (EDL).

EDIT DECISION LIST

Scene	Tape No.	Shot No.	Edit Type	Transition	Time code/counter numbers		Duration	Description
					In	Out		
					(tape no. is designated hour)			
	6		AI	FADE UP	00:21:23:25	00:21:25:22	01:27	MUSIC
23	7	23–3	V	FADE UP	02:10:35:10	02:10:36:20	:30	CU2SHOT
24	7	24–7	V	C	02:18:22:11	02:18:23:26	:45	WS

Figure 4-12 An edit decision list (EDL) includes audio as well as video. Note on this example that the audio continues through more than one video edit.

1. If the action is with the same people in the same place at the same time from one shot to the next, the lighting should not change drastically, the actors should have on the same clothes, and the set and props should remain the same. This may sound simple, but there are horror stories, even in the broadcast business, of shows that let something simple ruin an entire show. In short, keep things consistent from one shot to the next. See Figure 4–13.

2. Maintaining continuity of screen direction is also important. If the actor is walking toward a car in one shot, when we cut to another angle, the actor should still be walking toward the car even if we cannot see the car. Screen direction also goes a bit further than this. If the actor is walking toward frame left in one shot, it would be awkward to cut to a shot of the same actor walking toward frame right. The change of direction would throw the viewer. If you need to have the actor change direction, you can insert a neutral shot like a CU or head-on shot in between to ease the transition. See Figure 4–14.

3. Screen position is also important. If an actor is sitting in one shot and the next shot has him standing, it could be awkward for the viewer, not to mention downright distracting. See Figure 4–15.

4. Matching action is another criteria of good editing. If a performer, for example, is exiting a house, you want to get her out of the house without "jump cutting" from one side of the door to the other. How would you edit this together? You could use CUs of the door being opened as a transition between one side of the door and the other. Or you could use CUs of the performer. Or you could just follow her through, cutting ultimately to the other side of the door. It could be done many different ways, but the important thing to remember is that you cannot just cut her from one place to other. You have to get her there. See Figure 4–16.

The idea is to think about the shots not only as they relate to telling the story, but as they cut together. The bottom line is always: Do they work?

Pacing and Timing

An editor's most profound influence on a production can be seen in the pacing and timing of shots. In other words, how long does a shot last and where does the shot go inside the program and how long does the transition last? Based on other edits, an editor could elect to make a shot longer or shorter to emphasize or de-emphasize some aspect of the program. A quick kiss, for example, could become a long, passionate kiss, depending on the duration of the shot. The length of the shot could totally change the viewers's reaction to the scene.

An editor could change the pacing of a scene by the way it is edited. The

(a)   (b)

(c)

Figure 4-13 Continuity of lighting must be maintained within a scene. Of these three shots, a and b could be cut together even though the lighting has minor changes, but shot c could not be cut from either a or b. The lighting changes too dramatically.

"walk-walk-walk" concept of real-time video production can be replaced with a more entertaining abbreviated version of the scene. For example, let's say a person needs to get from one location to another and the audience knows the distance between the two places. The natural pacing would be to let the person move from location to location in real time. The editor could elect to depart from this by cutting from a wide shot of the person to a closeup and then back to a wide shot, moving him forward to his destination at a faster rate. By cutting away to the CU, the editor has diminished the time needed to make the move and thus tightened the pacing. See Figure 4–17.

A final aspect of timing is related to transitions. There is no time variable on a cut, but there is one on a dissolve and a wipe. How long does it take to do a dissolve from one shot to another, for example? A dissolve can last ten frames or 2 seconds. A wipe has the same time variable. How long a dissolve or wipe lasts

(a)

(b)

Figure 4-14 Continuity of screen direction means maintaining the direction of the actor as well as maintaining the background. These two shots will not cut together. The actor's direction is consistent, but the background changes.

is determined during the edit, based on what fits best into the design of the program.

Pacing and timing keep the action moving in a logical, sequential manner. An editor can put his or her mark on a show by the pacing and timing. The movie "Star Wars" is a classic example of effective pacing and timing, particularly during the fighting sequences in space. Cuts from the cockpit of one ship to the cockpit of another to exteriors of ships maneuvering to ships blowing up, all intercut with voice-over cockpit conversation, put the viewer right into the action. These scenes were made in the edit, not during shooting.

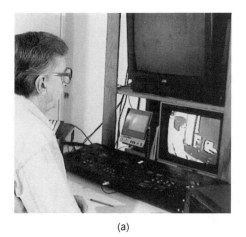 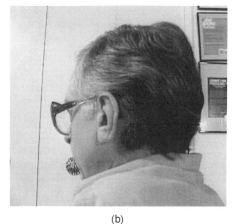

(a)   (b)

Figure 4-15   Screen position should not change from one shot to the next. If the actor is looking screen left in one shot, he should be looking screen left in the next shot.

GETTING FROM ONE SHOT TO THE NEXT

Getting from one shot to the next is called a *transition*. There are three basic transitions from which to choose: cut, dissolve, and wipe. Variations on these are the fade and jump cut. Not all editing systems can perform all these transitions, so the first thing to do is to find out what your equipment can do. It is pointless to plan to do a transition that your equipment has no ability to execute.

Cut

The most used transition is the cut. Even the most basic equipment can do a cut, which is simply going from one shot to another abruptly. An editor can perform a video-only cut or an audio-only cut or cut both simultaneously. See Figure 4–18.

Jump Cut

A variation on the cut is the jump cut. A jump cut moves a person or object from one location to another abruptly, for example, moving a person from inside the house to outside the house. If a jump cut is not done for effect, it is a mistake to do one because it is jarring to the viewer and may be perceived as a mistake. On the other hand, a jump cut can be extremely effective if planned as part of the production. It can be an excellent comic effect, with the performer jumping from

(a)

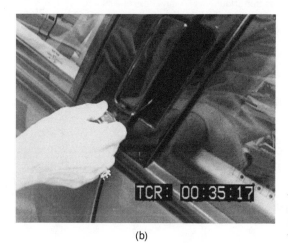

(b)

Figure 4-16 Action should match from one shot to the next. These two shots will not cut together. The actor's left hand is on the door handle in the wide shot but not there in the closeup.

scene to scene, or it can be used to establish the presence of a supernatural being, such as a ghost. To avoid using it incorrectly, it is best to think of a jump cut as an effect, not a transition. See Figure 4–19.

Dissolve

The second most used transition is the dissolve. A dissolve gradually fades out one image, while simultaneously fading in another. In a sense, the two images pass each other, melting into each other at the height of the dissolve. A dissolve

(a)

(b)

(c)

Figure 4-17 Pacing can be tightened by cutting from a wide shot to a closeup and then to a wide shot, condensing the time it takes for the action to happen in real time.

may be used to soften the transition between two shots. For example, if a person is outside the house in one shot and, in the next shot, he is inside talking on the telephone, an editor may elect to do a dissolve between the shots so that the person is not jump cut inside, and the viewer accepts that it took some amount of time for the person to get inside and pick up the phone. See Figure 4-20.

Fade

A variation on the dissolve is the fade, which is the gradual increase or decrease of the video image or the audio. Video images generally fade in or fade out from black or another color, while audio fades up and down. Most programs begin with

(a)

(b)

Figure 4-18   The most common transition is a CUT, which is simply going from one shot to another abruptly. Note the time codes on the shots. The wide shot ends on frame :02, cutting to the closeup on the next frame, frame :03.

a fade in from black. Most equipment is capable of doing a fade in or out. See Figure 4–21.

Wipe

A wipe is a transition that requires equipment with minimal special effects capability. When doing a wipe, one picture is literally wiped off as another is wiped on. It is as if one picture moves over another, covering it up. Wipes can usually be done in a variety of patterns. For example, the picture can be wiped in on a diag-

(a)

(b)

Figure 4-19 A jump cut moves a person or object without a transition. For example, if photo (a) cuts to photo (b), the person would jump into the scene.

onal, from left to right, or in from the center in an expanding circle. See Figure 4-22.

PLANNING DELIVERY

Broadcast television programs are delivered in a certain format that gives the tape operator standard cues for playback. This works with a timing sheet to provide complete playback information. Once the program is edited together, you

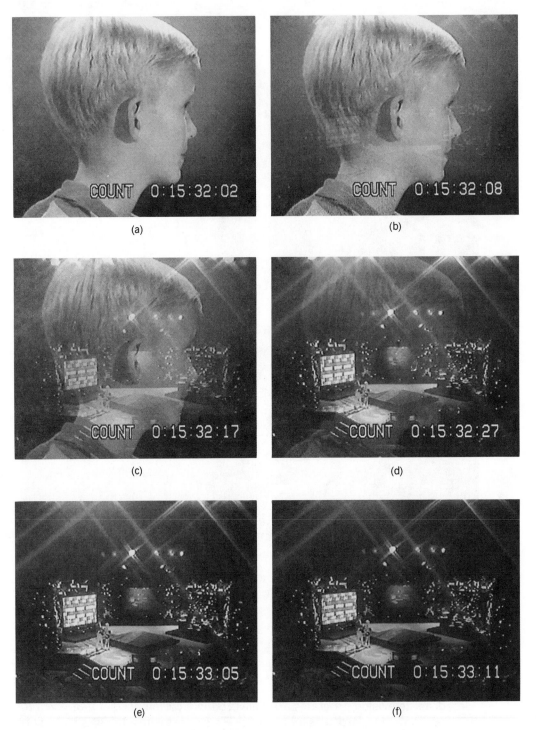

Figure 4-20 A dissolve gradually fades out one image while simultaneously fading in another. In this series of photos, the dissolve begins at 0:15:32:02 and ends at 0:15:33:11, lasting :01:09, or 1 second, 9 frames, or 39 frames.

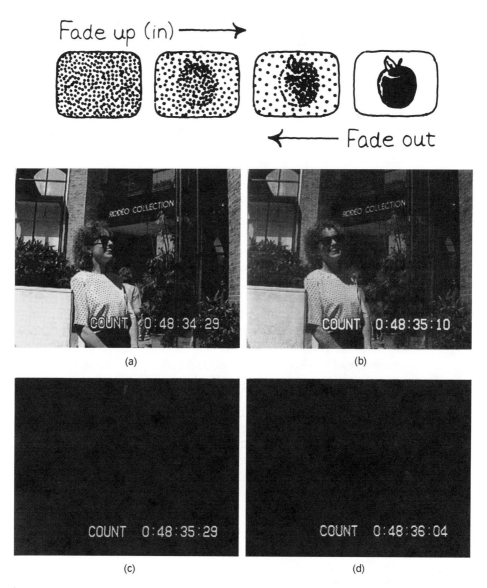

Figure 4-21 A fade is a gradual increase or decrease of the video image or the audio. Most programs begin with a fade up from black and end with a fade out to black. In this series of photos, the picture fades to black in 1 second, 5 frames or 35 frames, beginning at 0:48:34:29 and ending at 0:48:36:04. (James R. Caruso/Mavis E. Arthur, A BEGINNER'S GUIDE TO PRODUCING TV© 1990, p. 24. Reprinted by permission of Prentice Hall, Englewood Cliffs, New Jersey.)

need to do several things to finish the post-production process: (1) duplicate the edit master for delivery, and (2) make the timing sheet.

Delivery Recorded Format

Before you begin to edit, insert the edit master into the record machine and rewind to the beginning of the tape. Then follow these steps:

1. Record at least 15 seconds of black so that there is enough control track for the tape to get up to speed.
2. Record at least 1 minute of color bars and a reference audio tone. With these the tape operator can set up the video and audio.
3. Record approximately 10 seconds of program slate that identifies the program, as shown in Figure 4–23.
4. Record 10 seconds of video black, if the tape is not already blacked. If the tape is black, play in 10 seconds. If you want to put a visual countdown on your program, this is the place to put it. Instead of video black, record 1 second each of numbers 10, 9, 8, 7, 6, 5, 4, and 3. Seconds 2 and 1 should be video black so that they do not interfere with the program itself.
5. Record the program material. This segment includes all acts, as well as commercial black or commercial inserts. If the TV station is to insert its own commercials, you record video black in place of the commercial. If the commercial is yours, edit it in the appropriate place. Black commercial pods should be the exact length of the commercial pod; that is, if it is a 2-minute pod (four 30-second commercials), there should be 2 minutes of video black. With the tape delivered this way, the tape operator will not have to stop and recue the tape from the beginning to the end of the program.
6. Record 10 seconds of video black after the show ends.

Timing Sheet

A timing sheet is a courtesy to your viewers. It outlines the exact timing of the program. All broadcast shows provide these to television stations to let them know exactly how long every act runs, how much commercial time is used by the distributor or network, and how much time is available to the local station. TV stations also use these timing sheets to prepare their daily logs, which map out the entire broadcast day.

A timing sheet, as shown in Figure 4–24 and included in the Appendix for copying purposes, is laid out to accommodate either a half-hour or an hour program, with commercial insertions at the appropriate places. All broadcast shows stick pretty much to this format.

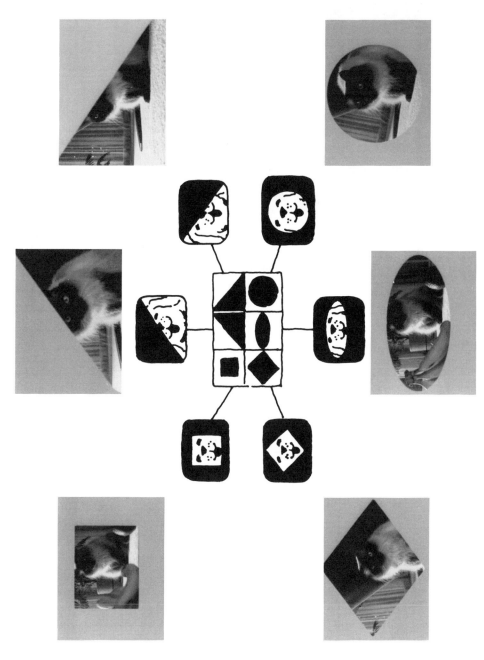

Figure 4-22 A picture can be wiped on or off, usually in a variety of patterns as shown here. (James R. Caruso/Mavis E. Arthur, A BEGINNER'S GUIDE TO PRODUCING TV© 1990, p. 24. Reprinted by permission of Prentice Hall, Englewood Cliffs, New Jersey.)

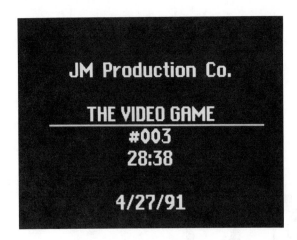

Figure 4-23 A program slate recorded before you begin editing identifies the program. The one shown here includes the production company, the name of the show, the show number, the total run time (this includes three 2-minute commercial breaks), and the taping date.

Note that there is place to indicate whether the tape is recorded drop frame or nondrop frame. As we have learned, this makes a big difference in how long the program will actually play in real time.

Distribution Dubs

You never want to play the program back on the edit master if it is going to be played more than once, and maybe not even then. The edit master, after all, is your first copy. It represents the best technical quality. If you make no dubs, it is also the only copy of the program. If you should damage it, the program may be damaged forever. If you lose it, there is no program at all.

The best solution is to make a dub of the edit master for playback in all cases. Protect the edit master by shelving it. Bring it out only to make additional dubs. Never play it for your viewers.

AN EDIT RULE

There is an unspoken rule in well-produced television that no camera shot should last longer than 30 seconds and that no single scene should last longer than 3 minutes. We call it the 30–3 rule. The reasoning is that the audience becomes bored rapidly and that even the slowest program will appear to have more action if the shots and scenes continue to change at this rather rapid pace.

A 30–3 philosophy keeps the program moving and the viewer interested. The viewer has been trained by the 30–3 rule. It is a fact you must face head on, adhering to these unspoken guidelines for creating watchable television.

TIMING SHEET AND CERTIFICATE OF PERFORMANCE

AIR DATE _____ THE VIDEO GAME   STATION _____

SHOW NUMBER 003   DROP FRAME _____   TIME 29:58 INC.
                    NON-DROP FRAME XXXX   COMMERCIALS

		HOURS	MIN	SEC	FRAMES	COMMERCIAL NAME/CODE	NOTE
COMMERCIAL 1		01	00	00	00	DR. PEPPER JUNGLEMAN	
ACT I	IN	01	00	30	00	Includes :15 Billboard	
	OUT	01	07	23	00		
COMMERCIAL 2		01	07	24	00	SEGA "FOREIGNER"	
3		01	07	54	00	HERSHEY "COUSIN WILLIE"	
4		01	08	24	00	BLUTH ANIMATION	
5		01	08	54	00	KIDS FOR KIDS "OWENS"	
ACT II	IN	01	09	25	00		
	OUT	01	18	20	00		
COMMERCIAL 6		01	18	21	00	DISNEY "RAINBOW BRIGHT"	
7		01	18	51	00	"	
8		01	19	21	00	"	
9		01	19	51	00	"	
ACT III	IN	01	20	22	00		
	OUT	01	23	58	00		
COMMERCIAL 10		01	23	59	00	SEGA "A FUNNY THING"	
11		01	24	29	00	HERSHEY "COUSIN WILLIE"	
12		01	24	59	00	MICHAEL JACKSON PSA	
13	xxxxxxxxxxxxxxxxxxxxxxxxxxxxxxxxxxxxxxxxxxxxxxxxxxxxxxxxxxxxxxxxxxxxxx						
ACT IV	IN	01	25	30	00		
	OUT	01	29	08	00	Includes :15 Billboard	
COMMERCIAL 14		01	29	09	00	HERSHEY "COUSIN WILLIE"	
TO BLACK		01	29	39	00		

Figure 4-24 A timing sheet indicates length of acts, commercial pods, and total length of show.

CONCLUSION

Post production is the final phase in the creation of a television program. It includes all the activities that follow the completion of the taping up to the finished program, including, but not limited to, the selection of shots and deciding on the transitions between those shots; the selection of music, audio, and sound effects; the selection of graphics or other visual elements; and, finally, the execution of these decision in an edit session.

Three planning steps are absolutely essential to successful editing. These are the concept, shot logs, and EDL. The concept gives you direction, shot logs tell you what you have on tape, and the EDL outlines how you have decided to put the program together.

An EDL is the paper edit of the program. It is the step by step list of everything that will be in the program. A good EDL not only forces you to make final decisions by helping you see it in your mind, but it also saves you time, thus making your edit session easier. A good EDL frees you from making the basic decisions during the edit session itself and allows you to think about the extras you could add to make it even better.

It is an editor's job to make sure the edits work together, not to just cut the program. Ultimately, it is up to the editor to exclude shots that do not work and put in shots that will. Major considerations for the editor are continuity, pacing, and timing. Continuity dictates that the editor never cut two shots together that, because of their content or framing, distract from the program. The pacing of the shots, particularly how long they last, has an impact on the total production. Timing is also of importance, both as to where shots are edited into the program and how long transitions last.

There are three basic transitions from which to choose: cut, dissolve, and wipe. Variations on these are the fade and jump cut. Of the three, a cut is the most used, with all equipment able to execute this transition. A dissolve is the second most used, and a wipe requires equipment with special effects ability. Most programs begin with a fade up from black and end with a fade out to black. If it is used improperly, a jump cut can be perceived as a break in continuity and thus it is best thought of as a special effect.

The one rule that an editor should always keep in mind is the 30–3 rule. The viewer has been trained by this rule and expects no less.

TO DO

Prepare an EDL working with the opening scenes of "Who Shot Jack?" The program has already been taped and the shot logs for these opening scenes are provided for your use in making the EDL. Normally, you would have been at the taping and would have a feel for what the shots look like, but let's assume for the sake of this exercise that you were not available on these particular taping days.

While preparing your EDL, keep the following in mind:

1. Programs are not necessarily shot in sequence, but rather when it is most convenient in terms of the location and the performers. *Bottom line:* Shots on the log are not shot in sequence.

2. Just because a shot has been recorded does not mean it will necessarily be used in the program. A good director will shoot extra shots to be used if he or she needs a cover shot or a cutaway. Also, a particular shot may be taped several different ways to give the producer some options during the edit. *Bottom line:* Not all shots on the shot logs will be used, and there may be more than one version of a shot, so you can choose the one you think most effective.

3. Sound effects are usually taken from a sound library and added in post. They are generally not taped in the field. *Bottom line:* Sound effects are not on the shot logs, but you are still expected to include them on the EDL, indicating a sound library as the source.

4. The shot logs are not necessarily just the way they would be if made during the actual taping. We have taken some liberties to make them clear and excluded repeated takes for the sake of brevity. Most importantly, time code numbers noted on the shot logs have been shortened from what they might normally be. For example, we may really record 10 minutes of a WS, but in reality only 3 minutes of it is usable. For the purpose of this exercise, you may assume that every frame of a shot as logged is good enough to be used in the program. For example, if the shot log says you have 3 minutes of a WS of the exterior of the Regis Mansion, you have 3 clean minutes you can use.

5. Remember that you do not have to use every second you have recorded. You can use only 1 second of the WS Regis Mansion. Also remember that you can use part of a shot and then go to another shot and then return again to the first shot. For example, you can open on a WS of the Regis Mansion, cut to a CU of a tree blowing in the wind, and then cut back to the WS of the Regis Mansion. These are edit decisions, based on how you see the scenes unfold in your mind.

With all this in mind, make a copy of the EDL form provided in the Appendix or make your own on a piece of paper and decide how you would begin this dramatic program. Edit this opening to approximately 03:30 ($3\frac{1}{2}$ minutes). Do not forget to include music and sound effects on the EDL. Note sound effects coming from SL (sound library).

And now, here's the opening scenes of "Who Shot Jack?" The shot logs can be found in Figure 4–25. A suggested EDL can be found in Figure 4–26.

Who Shot Jack?

The Script

Note: Numbers 1 through 8 indicate shot numbers. In some cases, we have shot more than one version or several shots as a part of one of these base shots and have indicated this by numbering them consecutively, with the base shot the first number; For example, 1-1, 1-2, and 1-3 are all part of shot 1.

1.   OPEN ON EXTERIOR SHOT OF THE REGIS MANSION. IT IS HUGE, OMINOUS, AND COLD. IT IS EARLY EVENING AND THERE IS NO LIGHT COMING FROM ANY OF THE WINDOWS. THE TREES SURROUNDING THE MANSION BLOW VIO-LENTLY IN A HIGH WIND AND RAIN THREATENS TO BEGIN AT ANY TIME. SUDDENLY LIGHTNING STREAKS ACROSS THE SKY.

SOUND: THUNDER

PUSH TO WINDOW ON THE UPPER FLOOR.

2.   CUT TO INSIDE THE HOUSE, CONTINUING TO PUSH THROUGH THE ROOM, OUT INTO THE HALL AND DOWN A WINDING STAIRCASE INTO THE LIBRARY.

3.   CUT TO DIFFERENT ANGLE TO SHOW JACK REGIS SITTING IN AN OFFICE CHAIR, THE CHAIR TURNED AWAY FROM THE DOOR. HE SEEMS TO BE STARING OFF INTO SPACE.

SOUND: THUNDER

CU JACK'S FACE. THERE IS NO EXPRESSION ON IT. IT LOOKS LIFELESS.

4.   CUT TO EXTERIOR ... CAMERA IS IN FIRST PERSON RUN-NING THROUGH A WOODED AREA. IT IS ALMOST DARK.

SOUND: SOMEONE CRASHING THROUGH THE UNDER-BRUSH

THE PERSON FALLS.

SOUND: PERSON FALLING

5.   CUT TO PERSON ON THE GROUND. THE PERSON IS WEAR-ING A RAIN HAT AND COAT. IT IS IMPOSSIBLE TO SEE WHO IT IS OR TO DISTINGUISH WHETHER IT IS A MAN OR A WOMAN. THE PERSON LOOKS ANXIOUSLY OFF IN THE DIS-TANCE.

6.   CUT TO WIDE SHOT OF THE HOUSE. LIGHTNING CUTS A
      RAGGED PATTERN ACROSS THE SKY.

      <u>SOUND: THUNDER</u>

      <u>EFX: LIGHTNING</u>

7.   CUT BACK TO THE PERSON IN THE WOODS, HE/SHE GETS UP
      AND PLUNGES INTO THE BRUSH AGAIN, RUNNING AWAY
      FROM THE HOUSE LIKE THE DEVIL HIMSELF WERE AFTER
      HIM/HER.

8.   CUT TO WIDE SHOT OF THE HOUSE.

One More Thing to Do

First, did you notice in the suggested EDL that we used shot 1-1 three different
times, for shots 1, 6, and 8. No point in reshooting the same thing over again.
Also, we taped more than one shot for any one shot designated on the script,
giving the scene more life and excitement. A writer would not normally designate
every shot. This would be left up to the discretion of the director and he or she
would interpret the narrative into shots that develop the program concept.

   Using the suggested EDL provided here, determine how long the show will
be up to this point. Remember we asked you to keep this to approximately 03:30
($3\frac{1}{2}$ minutes). Once you have determined how long this EDL lasts, add up your
own. Did you keep within the suggested time frame? Did we?

   You will find the exact time of our EDL in brackets in the upper-right top
corner.

   Note: 00:00:00:00 denotes hours:minutes:seconds:frames. There are 30
frames to a second, 60 seconds to a minute, and 60 minutes to an hour. To determine the length of the shot edited in, subtract the out time from the in time. For
example,

$$
\begin{array}{r}
02:23:19:11 \\
-\,02:22:14:11 \\
\hline
00:01:05:00
\end{array}
$$
(1 minute, 5 seconds, zero frames)

Remember that the frame column reads up to 29 and changes to one frame on
30, just like seconds change to 1 minute when they count up to 60, that is,
00:00:29 to 00:01:00 and 00:59:00 to 01:00:00.

SHOT LOGS: WHO SHOT JACK?

Location: Exterior, Regis Mansion

Tape No.	Shot No.	Take No.	Time code/counter numbers In	Out	Audio	Notes
1	1-1	1	01:03:03:00	01:07:07:12	NO	EWS static
1	1-2	1	01:07:07:12	01:10:00:15	NO	EWS push to front door
1	1-3	1	01:10:00:15	01:20:14:23	NO	EWS push to upper window
1	1-4	1	01:21:15:16	01:23:24:10	NO	EWS truck to upper window
1	1-5	1	01:24:19:11	01:26:19:11	YES	CU tree blowing in wind
1	1-6	1	01:27:21:22	01:31:24:11	NO	WS night sky (lightning effect)
2	4-1	1	02:01:10:18	02:04:10:18	YES	POV person running through woods
2	4-2	1	02:05:00:19	02:08:30:23	YES	MS person running through woods
2	5-1	1	02:17:35:12	02:18:45:23	NO	CU face/obscured by hat
2	5-2	1	02:18:45:23	02:19:15:23	YES	POV person falling in woods
2	7-1	1	02:23:16:28	02:24:44:17	YES	MS person getting up from fall
2	7-2	1	02:24:55:19	02:25:45:13	NO	ECU eyes looking off
2	7-3	1	02:11:53:07	02:15:29:09	YES	WS person running away/woods

Figure 4-25 Use these shot logs for "Who Shot Jack?" to complete the TO DO exercise.

SHOT LOGS: WHO SHOT JACK?

Location: Interior, Regis House

Tape No.	Shot No.	Take No.	Time code/counter numbers		Audio	Notes
			In	Out		
3	2-1	1	03:07:37:15	03:09:55:27	NO	CU window, then pull and turn and truck out of the room into the hall, down a winding staircase, and into library where it stops on WS that reveals chair turned away from camera. We can see an arm hanging loosely over the arm of the chair. Truck around chair and zoom into Jack's face.
3	2-2	1	03:11:47:19	03:12:49:11	NO	CU window, then pull and turn and truck out of the room into the hall.
3	2-3	1	03:15:55:19	03:17:10:04	NO	POV going down a winding staircase and up to the library door.
3	2-4	1	03:17:18:12	03:18:38:11	NO	WS library revealing chair turned away from the camera. We see an arm hanging loosely over the arm of the chair.
3	3-1	1	03:27:19:01	03:28:19:01	NO	CU arm hanging loosely over the arm of the chair.
4	3-2	1	04:00:45:01	04:02:01:12	NO	MS chair, TRUCK around it to reveal Jack, pop zoom to CU of his face.
4	3-3	1	04:02:11:01	04:03:11:01	NO	CU Jack's face - static.
4	3-4	1	04:04:45:12	04:05:59:11	NO	CU Jack's face. His head falls forward.

Figure 4-25 (continued)

SUGGESTED EDL FOR OPENING SCENES OF "WHO SHOT JACK?"

This EDL may or may not be similar to the one you prepared. That does not mean that yours is wrong and this one is right. It means only that you have a different way of seeing it than we do. It also demonstrates what we have been saying: an editor has an impact on the program by the way it is edited.

EDIT DECISION LIST
[T:04:11]

Scene	Tape no.	Shot no.	Edit Type	Transition	Time code/counter numbers In	Out	Duration	Description
					(tape no. is designated hour)			
1	SL		A	FADE UP				THUNDER STORM under through edit #15
	1	1-1	V	FADE UP	01:03:03:00	01:03:13:00	10:00	EWS static house
	1	1-5	V	DISSOLVE	01:24:19:11	01:24:34:11	15:00	CU tree
	1	1-6	V	CUT	01:27:21:22	01:27:31:22	10:00	WS night sky
		EFX	V	KEY				Lightning over
	SL	SFX	A					Accompanies lightning
	1	1-4	V	CUT	01:21:15:16	01:21:45:16	30:00	EWS/truck window
2/3	SL		A	FADES UNDER				THUNDER STORM
	3	2-1	V	CUT	03:07:37:15	03:09:37:15	2:00:00	CU window/Jack in library

Figure 4-26 This is a suggested edit decision list for the opening scenes of "Who Shot Jack?" Use it as an example answer for the TO DO exercise.

Scene	Tape No.	Shot No.	Edit Type	Transition	Time code/counter numbers (tape no. is designated hour)		Duration	Description
					In	Out		
	SL		A	CUTS UP				STORM
	SL	SFX	A					THUNDER
4	2	4-1	A/V	CUT	02:01:10:18	02:01:25:18	15:00	POV woods
	2	5-1	V	CUT	02:17:35:12	02:17:40:12	05:00	CU face
5	2	5-2	A/V	CUT	02:16:45:23	02:16:50:23	05:00	POV falling, sweeten fall
	SL	SFX	A					
	2	7-1	A/V	CUT	02:23:16:28	02:23:21:28	05:00	Getting up
	2	7-2	A	CUT	02:24:55:19	02:24:58:19	03:00	ECU eyes
6	1	1-1	V	CUT	01:03:03:00	01:03:13:00	10:00	WS EXT house
	SL	SFX	A					THUNDER
		EFX	V	KEY				Lightning over
	4	3-4	V	CUT	04:04:45:12	04:04:48:12	03:00	CU Jack
7	2	7-3	A/V	CUT	02:11:53:07	02:12:03:07	10:00	WS person running
8	1	1-1	V	CUT	01:03:03:00	01:03:13:00	10:00	EWS house

Figure 4-26 (continued)

153

5

Equipment Basics

Video editing and signal-processing equipment is proliferating by leaps and bounds. The computer-based editing units that control the movement of the videotape and trigger the edit are becoming more available, and user-friendly software is making them easier to use. At the same time, the recording and rerecording of the video images are becoming more precise through advances in electronic technology and new, more sophisticated recording techniques. Digital effects and video processing are becoming more common each day and more affordable. With all these advances in equipment, the quality of the end product is improving, with some consumer and industrial video productions approaching the look of broadcast programs.

What does all this mean? It means that you now have more options for effects, transitions, graphics, and other glitz to put into your TV program than were available to you before. It is up to you to find out what your equipment can do so that you can make maximum use of its capabilities. In short, know your equipment.

By knowing your editing equipment, you can design your program to fit it and not paint yourself into a corner with an edit or effect that either cannot be done or will adversely affect the look of your production. The look, as well as the creative elements, have a great deal to do with how the audience perceives your efforts.

It is ultimately up to you to learn all the nuances of your equipment. To get you started in the right direction and give you some ideas of what to look for, a discussion of some basic equipment features follows.

USING THE RIGHT CONNECTORS AND CABLES

Before discussing the various terminal jacks found on the different pieces of equipment and what they do, let's look at the different cables and connectors used to connect them.

Basic Types

RCA Phono Type.   The most common type of jack on consumer video equipment is the composite video RCA phono jack. It carries both video and audio signals. A plug is at both ends of the cable, with the receptacle a part of the piece of equipment. See Figure 5–1.

The cable has two wire conductors. The first is in the center of the insulation and carries the electronic signal. It is soldered to the center post of the RCA phono jack. The second wire is insulated from the first and is braided around the insulation of the center wire. This wire serves as the ground and as a shield to outside electrical interference and is connected to the outside metal rim of the RCA phono jack.

DIN Type.   Component video format camcorders and VCR/VTRs have the component video DIN type of video jack. These jacks have two pairs of pins. See Figure 5–2.

The DIN jack has four wires and four pins connected as two pairs of wires. One wire in the first pair carries the chrominance signal (C) and the insulation is red. The second wire in the pair is the ground and is separated from the first. The second pair of wires is insulated from the first and carries the luminance information (Y). This wire has white insulation and again it is separated with the insulation.

Each wire is connected to a separate pin on the DIN connector. If you look in the end of the DIN connector with the rectangular plastic guide at the bottom, the luminance pin is at 2 o'clock. Four o'clock is luminance ground. Chrominance ground is 8 o'clock and 10 o'clock is chrominance.

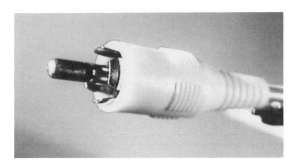

Figure 5-1   The most common type of jack on consumer video equipment is the composite video RCA phono jack, shown here.

Figure 5-2 A component video DIN jack, shown here, has separate wires that carry the chrominance signal (C) and the luminance information (Y).

BNC Type. The most common video signal connector used on industrial and broadcast equipment is the BNC connector (British naval connector). See Figure 5-3. It is used to carry video or a common sync signal. The receptacle part of the BNC connector attached to the cable has a positive lock twist barrel that secures it to two pins that protrude from the barrel of the jack on the video equipment. The cable's connections to the receptacle of the BNC are the same as for the RCA phono connector. See Figure 5-4.

F Type. The F-type connector is used to connect the 75-ohm round cable to the RF (antenna) in or out of the camcorder, TV, or VCR/VTR. See Figure 5-5. The center wire protrudes from the connector, and the outside barrel part of the connector, attached to the braided wire, is threaded to match the outside part of the jack. See Figure 5-6.

Connector Adapters. It is not necessary to have a cable for every possible combination of video-in/out terminals that might be found on the various pieces of video equipment, although it is certainly desirable. There are many different adapters that you can use that can be found in most electronic stores. So, if you have cables with the F-type connectors, you can get adapters to go from F to RCA, or BNC to RCA, or RCA to BNC, or almost any other combination that you might need, if you do not want to invest in a specific cable and connectors. See Figure 5-7.

Figure 5-3 The receptacle of a BNC type connector, shown here, has a positive twist-lock barrel that secures it to the connector on the video equipment.

Figure 5-4 The lock twist barrel on the receptacle of the BNC connector connects to the jack on the video equipment.

A word of caution about jacks, cables, and connectors. Video and audio signals are electrical currents that are adversely affected by a number of physical and atmospheric conditions. The most common effect is the loss of signal strength. The most common cause for this effect is dirt or corrosion on the jacks and cable connectors, so inspect and clean them regularly. Another cause of signal loss is cables that are too long, so keep them as short as possible in your editing setup.

Always use audio cables that are designed to carry the audio signal and video cables that are designed to carry the video signal. Do *not* use audio cables to hook up the video in your edit system. They will carry the video signal, but they will lose some of the signal along the way and it can affect the quality of your edit master. Do not use video cables to connect the audio terminals because the additional resistance will not let the full audio signal through to the next piece of equipment.

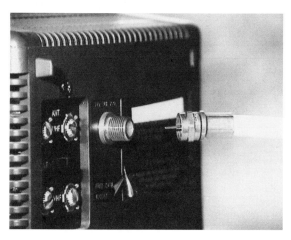

Figure 5-5 The F-type connector is used to connect the 75-ohm round cable to the RF (antenna) in or out of the camcorder, TV, or VCR/VTR.

Figure 5-6 On an F-type connector, the center wire protrudes beyond the threaded barrel.

VCR/VTR Hookups

A VCR (video cassette recorder)/VTR (video tape recorder) is primarily designed to record the images and sounds it receives through a camera. A camcorder is a camera and a video recorder in one housing. Some can record the images that are stored on various chips inside a camcorder, for example, the time, day, and date, or the chips that store titles or freeze frames, or the video for transitions like fades and wipes.

The cost of video recorders varies from a few hundred dollars for the home consumer, illustrated in Figure 5–8, to one that costs over $200,000 for a professional reel to reel VTR with all the bells and whistles, illustrated in Figure 5–9. Regardless of their size or cost, all VCR/VTRs record on the same medium, videotape.

The video signal that is recorded on the videotape is NTSC (National Television Standards Committee) in the United States or PAL or SECAM in other coun-

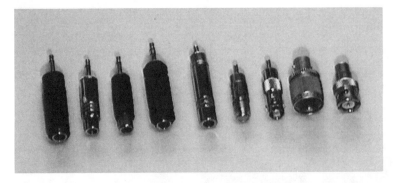

Figure 5-7 Connector adapters like these can be used instead of purchasing cables for every possible connection. Pictured are (left to right) a stereo phono to mini adapter, an RCA to mini mono adapter, a different style of RCA to mini mono adapter, a phono stereo to mono stereo adapter, a mini to RCA phono adapter, an F type to RCA phono adapter, a BNC to RCA phono adapter, a UHF to BNC adapter, and a BNC receptacle to RCA receptacle adapter.

Figure 5-8 Consumer VCRs such as the one shown cost hundreds not thousands of dollars. (Photo courtesy of Panasonic Audio/Video Systems Group.)

tries. A video signal standard means all videotape formats record the same way, so even if you cannot play a 1-inch tape in a VHS machine, you can wire the two machines together and record from one to the other with no problem.

While the video signal is the same, the types of recording formats are not. VHS is different from Beta; S-VHS different from 8mm, $\frac{3}{4}$ inch different from 1 inch. With only a few exceptions, you cannot play a tape from one format in

Figure 5-9 Professional video recorders are designed in various formats, including those that play and record on $\frac{3}{4}$-inch cassettes and on 1-inch reel to reel tape. (Photo courtesy of Ampex Corporation.)

another. For example, a VHS cassette will not even fit into a Beta machine, and if it did, it would not play. The exceptions to this rule are as follows:

1. VHS will play on S-VHS.
2. 8mm will play on Hi-8.
3. Beta, in some cases, will play on ED Beta.

In short, thanks to NTSC, when you go to edit, you can access video from any kind of recorded format, 1 inch, $\frac{3}{4}$ inch, VHS, Beta, or 8mm. All you have to do is cable together the different playback systems. Following are a number of different terminals on cameras, camcorders, and VCR/VTRs that allow access to the video signal.

RF Terminals.   All consumer camcorders and VCRs have a RF-out terminal (F-type) to allow you to hook the unit up to a television set. See Figure 5–10. The TV has an antenna terminal marked VHF or UHF that receives the broadcast signal over the air or through a cable system. See Figure 5–11. The broadcast signal is a radio frequency (RF) that is assigned to carry television broadcast signals and is the one that comes over the air, so your TV can receive the various broadcast stations in your area through your TV's or VCR's tuner.

The RF-out terminal is sometimes marked VHF or RFU out and is connected to the antenna (VHF)-in terminal of your TV set. The RF signal is only used for viewing or recording off the air and *never* for video editing. Although it is possible to dub the RF signal from a player to the RF signal of a recorder, it is not the most desirable way. The RF and video/audio signals are *not* the same and cannot be connected together.

If your TV only has antenna terminals that accept 300-ohm flat twin lead

Figure 5–10   All VCRs have a built-in F-type connector OUT to wire the unit to a TV. This OUT is marked ANT OUT, RF OUT, or VIDEO OUT.

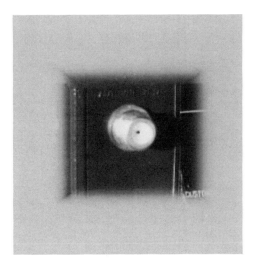

Figure 5–11 TVs have a video-in connector marked VHF/UHF that receives broadcast signals over the air or through a cable system.

antenna wire, it is necessary to connect these terminals to a matching box so that they match the camcorder 75-ohm round antenna cable and F-type connector. Some TVs come equipped with both the flat antenna terminals and the F-type antenna terminal. See Figure 5–12 for an example of a flat antenna wire, round cable with F connector, and a matching box. For playback with this RF connection, it may be necessary to set the switch located on the matching box and the TV to either channel 3 or 4.

Some camcorder manufacturers make their units with a RF in terminal so that you can use it as a VCR to record the broadcast signal through the TV's tuner. The RF terminal is generally built into the ac power and battery charger box. The video/audio-out terminals might be part of this box, and they are connected to video/audio-in terminals on the camcorder. See Figure 5–13. The

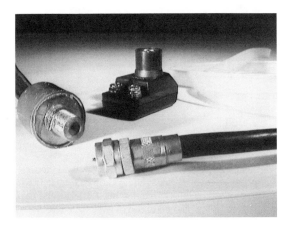

Figure 5–12 Pictured are a flat antenna wire with matching converter box, a round cable with F-type connector, and a matching box.

Figure 5-13 Some camcorders have an RF terminal built into the ac power and battery charger. In the one shown, note the four terminals on the right of the unit. One of these is for video in, one for video out, one for audio in, and one for audio out.

F-type terminal is the same RF connector that is included on all camcorders and VCRs so that the recorded videotape can be viewed on regular TVs.

VCRs have RF in/out terminals except for some recorders that are used just for professional recording or playback. This way the broadcast signal can be received by both the recorder and the TV. By connecting the cable system or an antenna to the antenna in terminal on the VCR and the antenna out to the antenna-in terminal of the TV, each can receive the broadcast signal for viewing and/or recording separate programs.

Video in. This terminal connects the video signal to the equipment. Any video-in terminal on a VCR or camcorder can be used to record a video signal from any video source. See Figure 5–14.

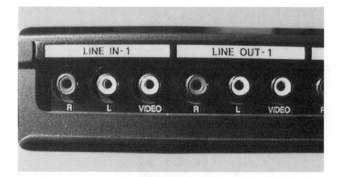

Figure 5-14 A video-in terminal brings video into a VCR from any video source. On the VCR shown, note that the video in is under the heading Line In-1. The R (right) and L (left) hookups are audio. Since the In is a number 1, there must be a number 2 In on this VCR, meaning it can be wired to two different video sources.

Video out.   This terminal sends the video signal out to any other piece of equipment with a video in. Whatever goes in must come out if you want to edit, dub, process, or just look at the video that you have recorded. See Figure 5–15.

Component Video DIN Terminal.   This video-in/out terminal is used with the separated signal of component video. It is necessary to have four-pin, DIN-type jacks and connector cables to carry the separated signal. See Figure 5–16.

Camera Terminal.   The camera terminal sometimes is a multipin terminal that connects the cable from a video camera to the VCR/VTR. See Figure 5–17. The camera cable has a positive locking threaded collar that is part of the jack that mates with the terminal on the VCR/VTR. The pins are connected to separate wires inside the camera cable.

The number of pins and wires depends on the manufacturer and the design of the equipment, its intended use, and when it was made. Some VCR/VTRs have a camera terminal that is simply a video in. The type of video camera used with this terminal has its own power supply, batteries, and/or ac and might or might not have an attached MIC.

Remote Terminal.   Used for cameras or VCR/VTRs that have wired remote controls, this terminal may be used on simple edit controllers to pause the machine when it is used as the player. See Figure 5–18.

Special Edit Controller Hookups

Edit Controller Terminal.   Through this terminal and its connector flows information from the player and recorder to the edit controller, which then

Figure 5-15 A video out sends the signal to a video receiver. Pictured is the video line out on a TV/monitor.

Figure 5-16 A component video DIN terminal accepts the separated signal of component video. Shown is this terminal on a VCR.

passes the commands back to the player and the recorder. The amount of information that flows back and forth to the recorder, player, and edit controller depends on three factors.

1. The capability of the VCR/VTRs to activate their own dynamics, that is, the method used by their tape transport to move and position the videotape to play and record simultaneously at an exact frame in the respective machines

2. The capacity of the edit controller to command the various functions of the

Figure 5-17 The camera terminal is a multipin connector on the VCR unit, as shown here.

Figure 5-18 A remote-control in and out pause terminal is used to pause a machine when it is used as the play machine in an edit setup.

machines that are necessary to correctly position the tapes to make the edits

3. The source of the edit information, which can be the VCR/VTR's tape counters, control track, time code, or some other electronic impulse encoded on the original and edit master videotapes

Some VCR/VTRs have an edit controller as part of their design and may be used as the recorder in an editing system. This type will have a terminal to connect the player with the recorder.

There are two different types of edit controller terminals. The first is a multipin jack that is unique to the particular manufacturer. See Figure 5-19. The second type is a single-pin terminal that accepts a mini phone jack. See Figure 5-20. Others only use one type of jack. The type of jack is determined by the degree of editing capabilities that a particular unit is designed to do.

Infrared Edit Controller Connector. Some manufacturers of basic edit controllers use the infrared remote control on VCRs in addition to or in place of a connecting cable to control the videotape movement and the other commands to the player and/or recorder. In place of the regular hand-held remote controller, there is a small wand that is placed in front of the remote control window on the VCR and is connected with a cable to the edit controller. With this type, it is necessary for the edit controller to learn the correct infrared commands to tell the particular VCR what to do. After the commands are learned, they are transmitted from the controller through the wand to the VCR. See Figure 5-21.

Figure 5-19 This is a multipin jack for an edit controller.

Audio Hookups

Audio in. This connection brings the audio signal into the piece of equipment. See Figure 5-22. Do not confuse the audio in with mic in, which is strictly a microphone input. An audio is used to input the audio signal from various sources, including those from another videotape, an audiotape, a record, or a laser disc source. If the equipment is stereo, it will have a separate audio-in terminal for each audio channel.

Audio out. This connection sends the audio signal to another piece of equipment. The receiving source can be a VCR/VTR, audio mixer, or any other piece of audio-processing equipment. If your video equipment is designed for stereo audio, it will have two audio-out terminals. See Figure 5-23.

Figure 5-20 Some VCRs with edit features have a single-pin terminal that accepts a mini phone jack, connecting it to an edit controller.

Figure 5-21 Some basic edit controllers use infrared remote control that uses a wand remote-sensing device connected by cable to the edit controller.

RCA Phono and XLR. The terminals on consumer equipment and the cable connectors are generally the RCA phono type with the two wires connected as described for the video cables. Some professional recorders use the RCA phono types of terminals and connectors also. Others use XLR types of connectors. The XLR type has three pins and is called balanced audio. See Figure 5-24.

With the various video terminals and connectors, it is not necessary to have audio cables with the exact connectors to match the audio ins and outs of your equipment. There are adapters for almost every combination of terminal and cable connector.

Audio Amplifier. If you are using a monitor to see the pictures and have the video out from the player connected to the video in of the monitor, you must also connect the audio out to the audio in of the monitor (if it has an audio amplifier) in order to process the audio signal so that you can hear the sounds that are

Figure 5-22 Audio terminal ins are designated either Audio In when only one channel can be input or R (right) and L (left) when two channels can be input.

Figure 5-23 Audio terminal outs are designated just like audio in terminals, either Audio In (when there is only one channel of audio) or L and R (when there are two channels of audio).

recorded on the videotape. TV–monitor combinations have an audio amplifier, so it is not necessary to connect the audio out to a separate amplifier.

However, if your monitor is not designed for audio, it is necessary to use a separate amplifier and speaker system. In this case the audio out from the player is connected to the audio in of the audio amplifier.

MIC in Terminal. Used to plug in a microphone to record audio on tape, this input can be found on cameras, camcorders, VCR/VTRs, and audio processors. See Figure 5-25. Whenever a separate microphone is connected to the

Figure 5-24 Some VCRs use XLR-type connectors. Pictured is a XLR connector receptacle (left) and XLR connector insert (right).

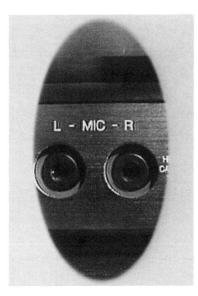

Figure 5-25 A mike in is used to input audio from a microphone. In the photo shown, the equipment can record the microphone on either the left channel of audio or the right channel or both.

MIC IN jack of most camcorders, the MIC built into the unit is made inoperative by disconnecting it, or on some camcorders it is automatically disconnected when the external MIC is connected. This prevents *two* separate audio signals from being recorded on *one* audio track.

If you want to record two microphones on one audio track, it is necessary to connect them to the MIC IN terminals of an audio mixer, with the mixed audio routed to the recorder, or use a Y connector. See Figure 5-26. A Y connector is an adapter used to mix audio signals. One end of it has two terminals, one for audio in 1 and one for audio in 2. The other end has only one audio out. The

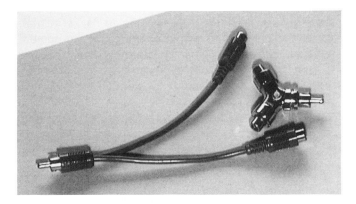

Figure 5-26 A Y connector is used to mix audio signals, combining an L and an R channel into one. Shown are two different types of Y connectors.

Figure 5-27 A phone jack is used to connect a microphone or a headphone.

stereo audio signal comes in the double connector as two separate signals and goes out through the other end as one audio signal.

Phone Jacks. Microphones and their cables are connected to most video recorders with a phone jack, not to be confused with an RCA phono jack. Others use the XLR jack. The phone jack has two wires isolated from one another in the barrel and tip of the jack assembly. See Figure 5–27.

If you have a microphone cable with a standard-size phone jack, adapters are available that have the larger opening for the standard-size connector on the receptacle end and the smaller size on the plug end. See Figure 5–28. The same is true if your VCR/VTR has an XLR MIC IN terminal. See Figure 5–29.

Headphone/Earphone Terminal. A headphone/earphone out allows you to hear the audio on the videotape through an earphone or headphones without amplification. See Figure 5–30.

HOW SWITCHES WORK

Switches turn equipment features on and off. Thus the number of switches has a direct relationship to the number of things the equipment is capable of doing. The more switches there are, the more things it can do. Switches come in a variety of sizes, shapes, and configurations. The first step of their operation is mechanical, with the finger touching, pushing, pulling, or sliding a button or some-

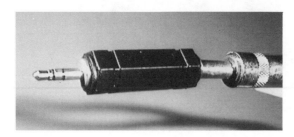

Figure 5-28 Phone jack adapters can be used to connect microphones.

Figure 5-29 XLR phono adapters can be used to connect a microphone to a VCR with an XLR mike-in terminal.

thing that looks like a dot, a recess, or a protrusion on the particular piece of video equipment. After that, the rest of the process is electronic. The switch turns an electronic circuit on or off or changes a function or the speed at which something happens inside the machine. Some of the things that switches do are obvious, like turning the power on or off to the piece of equipment, while others are not so obvious.

Let's look at the obvious first. See Figure 5–31.

VCR/VTR and Camcorder Switches

Play. This switch activates the videotape transport and turns on the video/audio play heads and the video/audio and RF out circuits. On some VCR/VTRs this switch is marked "forward" instead of "play."

Figure 5-30 A headphone/earphone out on a VCR allows you to hear the audio on the videotape through an earphone or through headphones without amplification.

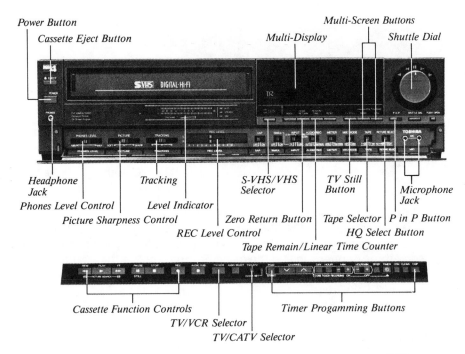

Figure 5-31 These are some of the switches you might find on your VCR.

Record. When the record switch is activated, it prepares the video-in and audio-in circuits to receive the signals and turns on the video/audio record and erase heads. There are two types of erase heads. One type is stationary and is mounted just before the drum that has the record head mounted on it. As the tape is pulled past the erase head, it is erased longitudinally, which means that it will leave part of the first few helical scan traces of any previous recording intact. See Figure 5-32.

The second type of erase head is mounted on the same drum as the record head. This a flying erase head. This type of erase head follows the identical path as the video record head and will erase the complete video trace prior to recording. If the VCR/VTR is the component format, it will also erase the hi-fi stereo audio trace. See Figure 5-33.

Fast Forward. This switch releases the tension that holds the videotape against the drum on which the heads are mounted and rapidly transports the tape to the take-up reel. A video recorder has mechanical guides that pull and hold the tape so that it can physically make contact with the drum. When fast forward is on, the guides release the tape on some tape transports so that it is completely retracted into the cassette if it is loaded in a cassette. Other types of

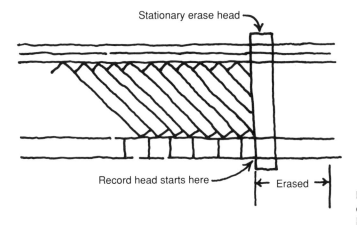

Stationary erase head

Record head starts here

← Erased →

View from Recording Side

Figure 5-32 A stationary erase head erases the tape longitudinally and will leave intact part of the first few helical scan traces of the previous recording.

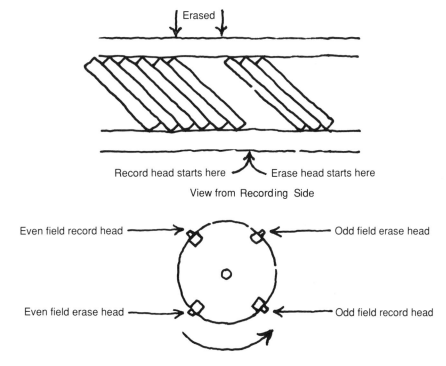

Erased

Record head starts here — Erase head starts here

View from Recording Side

Even field record head — Odd field erase head

Even field erase head — Odd field record head

Figure 5-33 A flying erase head follows the identical path of the video recording, erasing the complete video trace prior to recording the new video signal.

tape transports lightly hold the tape around the drum while in the fast-forward mode.

Rewind or Reverse.   This switch operates the same as the fast forward except that the videotape is transported to the supply reel.

Search.   Most VCR/VTRs can search forward or in reverse. The play head remains turned on and the videotape is in contact with the play head, which allows you to see the pictures in B/W or color depending on the features of the VCR/VTR. If the machine has special-effects play heads, the pictures will be stable in the search mode.

Stop.   This switch stops the tape transport mechanism and turns off all the heads in the drum.

Pause.   This switch stops the videotape but leaves it in contact with the spinning drum heads. On some VCR/VTRs the picture will be clear, while on others it will be jagged looking. Some VCR/VTRs rewind a few frames after the pause switch is activated, while others will rewind a few frames after it is deactivated before the play or recording starts. Because the videotape is in contact with the head drum while in the pause mode, most VCR/VTRs are designed to turn off the pause switch after a predetermined amount of time so that the videotape or the head will not be damaged by holding it in pause too long.

Jog/Shuttle Dial.   This dial combines most of the functions of the fast-forward, rewind, and pause switches when the play head is turned on. See Figure 5–34. It is located on the front of the VCR/VTR and/or on a remote control or on an edit controller. The outer part of the dial activates the shuttle mode, while the inner part handles the jog (frame by frame advance or reverse) function.

Figure 5–34   A job/shuttle dial combines the functions of the fast forward, rewind, and pause switches.

The jog/shuttle dial allows you to (1) combine several functions in a single knob, for example, search forward/reverse and frame by frame advance, and (2) search at a variable speed. Most jog/shuttle dials will play a steady, clear picture at all times except when in the fastest shuttle speed, in forward or reverse.

Videotape Recording Speed Switch. This can be a two- or three-position switch marked SP, LP, SLP, or some other speed designation. See Figure 5–35. This switch controls the velocity with which the videotape passes by the recording head. These tape speed markings can be found on most VHS and some 8mm VCRs and camcorders. Most professional VCRs and VTRs do not have this switch because they only have one speed, a speed designed for optimum recording quality.

Most VCRs, if they have three speeds, are marked as follows:

1. SP for *standard play (record):* This is the fastest videotape speed (velocity) for the machine. Set at this speed, a 120 tape will record for 2 hours.
2. LP for *long play (record):* This setting reduces the videotapes speed by one-half from the SP (fastest) setting and doubles the recording capacity on a given length of videotape. Set at this speed, a 120 tape will record for 4 hours.
3. SLP for *super long play (record)* or EP for *extended play (record):* This setting reduces the videotape speed to one-quarter of the SP speed and triples the recording capacity on a given length of videotape. Set at this speed, a 120 tape will record for 6 hours.

Beta VCRs use a different speed-switch marking system. They are designated as X-1, the fastest recording speed, industrial; X-2, standard consumer; and X-3, the slowest recording speed. It is not necessary to change the switch setting during playback to correspond to the speed that the videotape was originally recorded. The VCR will adjust internally.

Figure 5–35 The speed switch on a VCR determines the recording speed. In the photo, the VCR is capable of recording at SP or EP (sometimes called SLP) speed.

The *fastest* record speed is the *only one* that should ever be used in editing or recording original material to obtain the highest-quality recording. The video recording head is always spinning at exactly the same speed and tracing 30 frames of video information irrespective of the tape's velocity.

At the standard speed, a VHS videotape moves 1.31 inches per second. Beta videotape velocity is slightly faster at 1.57 inches per second. On $\frac{3}{4}$-inch machines the videotape moves 3.75 inches per second, while a type C, 1-inch professional VTR runs at 9.41 inches per second. Each is recording 30 video traces on their respective length of videotape. Beta and VHS formats used in professional or broadcast recording have a videotape speed approximately six times that of an industrial or consumer VCR.

Tracking. Sometimes the control track recorded on one VCR/VTR will not play back on another, which will cause the picture to be jagged at the top or bottom of the screen. The tracking adjustment to correct this problem is generally a small knob that turns left and right, with a center detent to locate the normal tracking. Some VCR/VTRs have two or three additional tracking adjustments for slow play or still play. Some make the tracking adjustment automatically.

Counter Reset. A counter is part of every VCR/VTR. Some counters can be indicators of the approximate location of a shot on the recorded videotape, while others will indicate a more exact location, depending on the type of counter. Some will give you the length of the recording in hours, minutes, and seconds. Some will even indicate frames, if not on the counter itself, on a display on the video monitor connected to the play VCR/VTR. There are four basic types of counters:

1. The mechanical type operates like a speedometer with three or four drums, each with numbers from 0 through 9 engraved on it. As the videotape is transported in either forward or rewind, the drums turn and indicate a higher or lower number as the videotape passes by a mechanical wheel connected to the drums. The numbers are arbitrary and are only relevant to the value that you assign to them and do not indicate recorded time. No numbers are recorded on the videotape.

2. The LCD or LED display type receives the videotape movement information from a mechanical or electronic sensor and gives it numbers displayed on an LCD or LED readout. It counts just like the mechanical counter.

3. The electronic counter type counts the control track pulses that are recorded on the videotape during the recording process. This type displays the information with an LCD or LED in hours, minutes, seconds, and sometimes frames of recorded time. If no control track is recorded on the videotape, the counter will not function because it has nothing to read. This type of counter is fairly accurate when compared to the previous counters since

it reads pulses recorded on the videotape, but it is not exact, particularly when the videotape has been fast forwarded, rewound, and paused several times.

4. The time code counter reads SMPTE time code encoded on the videotape. It reads like a 24-hour elapsed time clock in hours, minutes, seconds, and frames. Because the time code information is recorded on the videotape, it is *always* 100% accurate.

With the exception of the recorded time code counter, the reset switch will set the counter to zero. Some reset to zero automatically when a cassette is inserted, regardless of whether the tape is completely rewound or not. It your equipment has a mechanical counter that does not count in time code numbers, it is important for you to know how those counter numbers relate to real time. If you have not already determined this, use the instructions in Exhibit 5–1 to convert these counter numbers to real time.

Memory Switch.   The memory switch on a counter system is used to return to a specific place on the tape. For example, if you want to rewind to a specific shot, zero the counter at that location. Then switch the memory on. Later,

Exhibit 5–1 Counter Time to Real Time

If your counter numbers are arbitrary, then you must determine how many numbers change in 30 seconds of playing time. The following is one method of determining real time using the counter on the player and a stop watch. Use any videotape that has been recorded at the fastest speed, SP or standard play.

1. Cue the player to the desired edit-in point and press *Pause*. Write down the counter starting number. Let's say that the counter number is 0277.
2. Press *Play* on the VCR and the *Start* button on the stop watch simultaneously. Play the VCR for 30 seconds.
3. Press *Pause* on the VCR. Write down the ending number, for example, 0297.
4. Deduct the starting counter number from the ending number. 0297 minus 0277 equals 20 number changes in 30 seconds.

This calculation will help immensely when trying to edit with a punch and crunch system.

when you want to get back to that location, press fast forward or rewind (depending on whether the shot is forward or backward from your present location) and the tape will automatically stop when the counter reaches zero.

This is an excellent way to return to a specific spot, but by zeroing the counter in the middle of the tape, you lose the count from the beginning of the tape. This can make your shot logs or any notes you have made based on these numbers useless, since these numbers are generally based on zeroing the counter at the beginning of the tape. You will not find this switch on a machine equipped to read recorded time code. It does not need it. Time code numbers will take you to any point on the tape.

Index and Index Scan.   An electronic pulse is encoded on the videotape when this switch is pushed or the recording starts. To return to this index mark, push index scan. It operates just like a memory switch, except you do not have to zero the counter.

The number of indexes that the memory is capable of retaining is predetermined. Some VCRs automatically index each time the record switch is turned on. Most VCRs with this feature do not index when the record/pause mode is used. The same is true when the play/pause is used for index scan. The index mark can be erased by pressing the index erase button.

Tape Remaining.   This switch will tell you how much recording time is left on a video cassette, provided that the switch that tells the VCR/VTR the length of the videotape was set when the rewound tape was inserted. On some machines it is necessary to reset the counter to zero when you insert a new tape; on others, it happens automatically when a videotape is inserted.

Digital Effects.   The video signal is analog. When the digital circuit is turned on, it converts this analog signal to a digital signal. While the signal is in the digital mode, it can be manipulated. For example, the picture information can be frozen or flipped or turned or moved or you can use any of the sophisticated and complex moves that are seen on broadcast TV, depending, of course, on the capability of the digital device. Many pieces of video equipment made for consumer and industrial users have at least some digital features, such as PIP (picture in a picture).

Edit on/off.   This switch is used to turn on or off the line delay required for playback on a TV or monitor. This line delay is not needed for editing or dubbing videotape, so the circuit is turned off on the player.

Edit Controller Switches

Edit controllers are the heart of any editing system. Some are loaded with features, while others are little more than devices that start and stop the editing process in the recorder and the player. Regardless of their degree of sophistication, the controller's purpose is to relieve some of the mechanical steps in the total editing process, thus making it easier.

The edit controller console can be simple and basic, with the switches plainly marked to indicate what they do. See Figure 5–36. Other consoles can appear to be quite complicated because they resemble a computer keyboard. See Figure 5–37. In fact, most edit controllers are special-use computers that control a software program designed to start, stop, preview, and record the edit decisions that are made by the person doing the editing.

The software program is stored on disc or in the chips that are a part of the edit controller. The software program is the determining factor of the capacity of the controller's ability to perform certain functions and the accuracy at which it does them. The program can also be designed to do many things, including edit decision list management that allows the edit in and out to be changed to a different order or stored for future reference.

Editing accuracy depends on the accuracy of the edit controller *and* the VCR/VTRs that it is connected to. Most consumer and industrial edit controllers and recorders are considered accurate if they can edit within three frames. Professional and broadcast edits are to the exact frame.

The following are some of the edit controller switches, illustrated in Figure 5–38, and what they do.

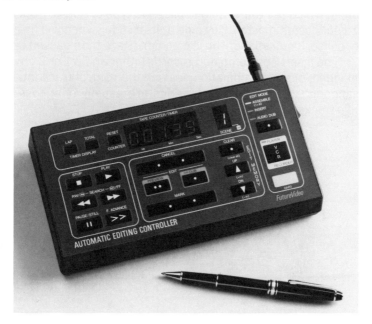

Figure 5-36 An edit controller can be simple and basic, with the switches plainly marked to indicate their function. (Photo courtesy of Future Video Products, Inc.)

Figure 5-37 Edit controllers are special use-computers. Some have keyboards similar to that seen on a computer. (Photo courtesy of Videonics)

Assemble and Insert Edit. Generally, two switches handle these commands, which tell the edit controller the type of edit the operator wishes to record on the edit master. On some units a slide switch will be moved from assemble to insert indications. These commands are then passed on to the recorder.

The insert edit command records either audio 1 or 2 or video only, or audio and video, or any combination, leaving the control track intact. If the VCRs are component video, insert audio edits are usually only possible on the normal audio tracks.

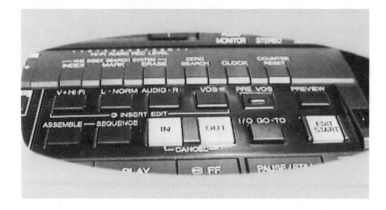

Figure 5-38 Shown is the editing switch setup on a high-end consumer VCR.

An assemble edit command tells the recorder to record all signals, including control track.

Most professional editing is done in the insert edit mode even though the shots are recorded sequentially. This preserves the continuous time code and control track recorded on a blacked edit master (BEM).

Mark. This switch tells the edit controller the edit-in/out points on the original material. Some basic edit controllers require the operator to mark these points "on the fly." Others will allow the mark to be initiated while the player is in pause at the exact in and out frames of the master material.

Edit controllers that use SMPTE time code will mark the edit in and out with the hour, minute, second, and frame when the editor identifies the exact location by typing in the time code numbers on a keypad, or it is done automatically when the mark button is pushed. In any method of indicating the information, the edit-in and edit-out points are stored in the edit controller awaiting the next command.

Some basic edit controllers will store in their memory multiple assemble edit-in and edit-out points for the player. By using this feature, the editor can make the decisions for several edits all at one time and store them in the edit controller as one single stream of editing.

Preview. This switch tells the player and the recorder to cue up and play or preview, not record, an edit. One edit can be previewed or the complete production, depending on the capacity of the edit controller to store the edit decisions.

Trim. If, after you preview the edit, you find the edit points are not exactly where you want them, you can add additional frames or subtract frames with the trim feature. For instance, if the edit-in point is too far into the shot, you can trim back to the exact place that you want it to start to record, for example, trim back five frames. You can also trim in.

Counter Reset. Some edit controllers allow you to reset or zero the counter of the VCR/VTRs attached to it by pushing this switch.

Fast Forward, Rewind, Search, Stop. Some edit controllers have the capacity to perform these functions for all VCR/VTRs connected. They duplicate the switches that are found on the record machines. Others have a jog/shuttle dial that will function the same as the one on the record machine.

Clear. This switch clears the selected in and out points. On some edit controllers the switch is marked *Back*, which will take the function back to the last command from the one it is presently on. In the mark mode, this would take the edit out back to the edit-in command so that it can be changed.

E to E (Electronics to Electronics). This allows you to see the player and the recorder's video in a single-monitor editing setup.

Input Select. This switch tells the recorder which input to record when there is more than one video input.

Lap Time. This switch keeps track of the length of the edit master in real time.

Record or Edit Start. This switch initiates the edit. Depending on the capacity of the control unit, it will also do several other functions, including the storage of the edit decision information on the edit management list.

Audio Switches

Video is video and there is only one video signal. Audio is something else. It's monaural, it's stereo, it's hi-fi, it's hi-fi stereo, and it's even digital. Audio may be recorded on one and/or two linear audio tracks and can be either stereo or monaural. On super hi-fi stereo formats the audio is recorded under the video information. Recording and mixing audio are often the *most* complicated parts of video editing. See Chapter Eight for a more complete discussion. For now, the following are some of the audio switches you may find on your equipment:

Audio Dub. This switch in its simplest form allows you to replace the previously recorded audio track on the videotape while leaving the video intact. It turns on the audio erase head and the audio record head. New audio, such as voice over (VO) or music, can be recorded on the linear audio track.

Audio Track. The switch marked audio track 1, 2, or both is used for playback only. On some VCR/VTRs the access to record is through the audio input jacks, while others have an audio-in switch that routes the audio signal to the selected track(s). If the VCR/VTR has two audio tracks, the individual tracks can generally be accessed for simultaneous playback of a VO on one track and a second effects track that can have music and/or mixed sound effects. The tracks can also be used to record two languages, different narration, or SMPTE time code recording.

Super format VCRs and camcorders record hi-fi stereo audio under the video track on the videotape. All the super formats, whether they are Hi-8, super VHS or ED, provide a second set of audio recording tracks and access switches. This is called normal audio and may be used to change the original audio recording.

Audio Level. The audio recording level is set by a slider-type switch that regulates the amount of audio signal that is being recorded on the videotape.

CONCLUSION

There is one simple fact that is true for everything you do in the video process: if you know what your equipment can do, you can make better and more exciting videos. Find out everything there is to know about your equipment, particularly before you begin to edit. While shooting can be creative with camera manipulation and the like, editing can also be a creative high. In the edit, the program is made or lost. And, yes, a good edit can save an otherwise mediocre program. By knowing your equipment, you have the best possible chance to make your editing easy and your program the best it can be.

TO DO

Using this chapter as a guide, look at your equipment—all your equipment, including your camera or camcorder, VCR/VTR, SEG, CG, and/or edit controller. Identify on each machine:

1. All terminals
2. All switches, both their location and function
3. All cable connector jacks and their mating terminals

Test any switch or function you do not understand. Spend some time learning your equipment. Find out what it can do, how you can connect it most effectively, and how you can utilize every feature on it. Use the owner's manual or any special training guides available to help you learn more about it.

6

Basic Editing System and Some Simple Editing Methods

The configurations of an editing system can be as different as the video productions that are edited on them. They can be as simple as wiring one VCR to another or as complicated as those that are found in state of the art post-production facilities with multiple tape sources and high-tech digital effects capabilities, for example, an edit controller, multiple playback and record VTRs, time base correctors, a video switcher with digital effects, waveform monitors and vectorscopes, video-processing amplifiers, a title camera, an audio board, audio speakers, and video monitors for preview and playback. See Figures 6–1 and 6–2.

The actual process of putting pictures together can be simple cuts only or it can be complicated with multivideo sources layered one over another, creating effects that, until recorded, exist only in the mind of their creator. The possibilities are limited by two things: (1) what the equipment is designed to do, and (2) the editor's knowledge and expertise at utilizing its capabilities. A limited editing system is just as challenging as a sophisticated one. Both test the editor's abilities to make it perform.

The basic editing system is capable of cuts only, as illustrated in Figure 6–3, and the simple editing methods described here are just that, basic and simple. However, this basic, simple approach is the basis for all video editing regardless of how sophisticated the video equipment appears to be. As you find out more about video editing, you will discover how valuable the basics are in learning how to do other transitions and effects within the video technology and its requirements.

So let's look at how you can hook up a basic editing system and then let's find out how to do some simple editing. Keep in mind that the basics are just

Figure 6-1   This is a basic edit setup, which includes a play VCR, a record VCR, a play monitor, a record monitor, a CG monitor, a CG, an SEG and an edit controller.

Figure 6-2   A full-blown state of the art post-production studio has multiple tape sources and high-tech digital-effects capabilities like this typical Ampex edit suite. (Photo courtesy of Ampex Corporation.)

Figure 6-3 The basic editing system is capable of cuts only. Note the time code numbers on this series of pictures. In (a), it reads 6 minutes, 14 seconds, 21 frames; in (b), 6 minutes, 14 seconds, 22 frames. The cut between these two pictures is made between frame 21 and frame 22. In (c) and (d), the cut is between frames 8 and 9. In (e) and (f), the cut is between frames 15 and 16 and, finally, in (g) and (h), the cut is between frames 27 and frame 28.

that, the basics. If a basic principle avoids a potential problem at the simple level of editing, it will do the same at the complex level when you have technically more advanced equipment and thus more capability.

BASIC SYSTEM

The basic edit system consists of two machines: a player and a recorder. It also includes one or two monitors or TVs and, if it is not built into your TV or TV/monitor, an audio amplifier and speakers. This basic setup is the backbone of all edit systems. In fact, this basic system was the only way to edit videotape as recently as the early 1960s. Prior to that, in the late 1950s, 2-inch quad videotape was cut and spliced the same way movie film is still edited. But today in TV we have

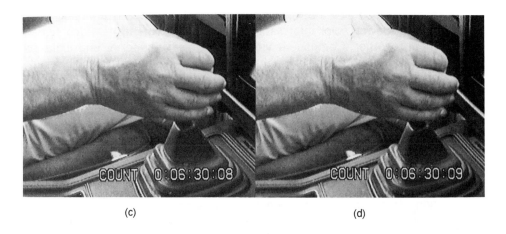

(c)   (d)

(g)   (h)

Figure 6-3   (continued)

magnetically encoded tape and electronic editing, and the cut and splice technique, at least for videotape, is a thing of the past.

As is true with most basics, if you understand the basic edit system and the simple editing steps needed to make it work, it will make it easier to understand more about the other pieces of equipment and what they do as they are added to this basic system. It will also make it possible for you to plan your production to fit the editing equipment that you have available. If there is an effect that your edit equipment is not capable of doing, do not plan to do it. It is that simple.

Before you begin, remember that the two machines in the basic edit system do not have to be the same format, but they must have compatible video signals. Both must be NTSC *or* PAL *or* SECAM. If the equipment was designed to be sold in the United States, it will be NTSC. As far as the formats are concerned, they can be any combination that you have available: VHS, S-VHS, VHS-C, SC-VHS, Beta, 8mm, HI-8, $\frac{3}{4}$ inch, or 1 inch.

This holds true for the formats recording a component video signal as well as those recording a composite video signal. A component video signal records luminance and color information separately. A composite video signal records them together. Component video is technically superior and looks better because it has more lines of resolution. Some current video recording systems that separate luminance and chrominance are labeled S-VHS, HI-8, and ED for *extended definition.* Most component video systems have input and output terminals connected through a DIN type-S connector that keep the signals separated.

You can connect the basic edit system to any of the component video formats and edit to a composite video format, though you will lose the advantage of the higher resolution that the component video system offers. If you want to keep the advantages of higher resolution, you must route the component signal, video out, to the next component video in to maintain the video signal separation. Most component video equipment is manufactured with dual video inputs and outputs to accommodate composite and component video signals.

Setting up

To set up the basic edit system, you need the following:

1. One machine (VCR or camcorder) to be designated as the player. The player is the machine that will control most of the steps in the simple editing methods described later.
2. One machine (VCR or camcorder) to be designated as the recorder. This machine will record the edit master.
3. Cables, either RCA phono type, DIN S-connector, F type, flat antenna wire, or BNC type.

4. One video monitor, TV set, or TV/monitor combination.

And, if you are not using a TV or TV/monitor with an audio amplifier built in:

5. One audio amplifier and speaker or headphones.

In all instances, in our descriptions of basic edit systems, VCR can mean camcorder. Monitor means that it does not have an audio amplifier or speakers. TV means that it does not have separate video/audio inputs and is only equipped with antenna-in terminals. The TV may be either black and white or color. See Figure 6–4.

Once hooked up, the video and audio signals will be sent from the player through the recorder's video and audio circuits, and you will be able to see and hear them on the TV or monitor as long as the recorder is turned on and is in the stop mode. When the recorder is in pause, you will see the still picture from the recorder and will not be able to see the pictures from the player. When the recorder is in the record mode, technically you will see on the monitor or TV both what is being recorded and what is playing on the play machine. In other words, this is what you will see on the TV or monitor using basic edit system 1 through 4:

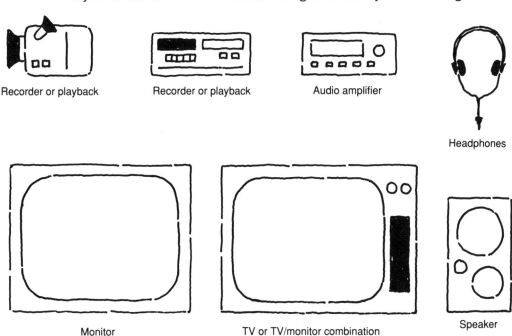

Recorder or playback   Recorder or playback   Audio amplifier

Headphones

Monitor   TV or TV/monitor combination   Speaker

Figure 6–4   The equipment included in the basic editing systems that follow include some of those shown here.

When recorder is on and in:	You see on the TV/Monitor the:	
	Record Machine	Play machine
Stop		See
Pause	See	
Play	See	
Record	See	See

There is a wide variety of VTRs, VCRs, camcorders, TVs, TV/monitor combinations, and monitors, all with different inputs and outputs. The following are several ways to hook up the basic edit system based on the equipment. Pick the one that matches the video equipment that you have.

Basic Edit System 1:   Use this setup if you have

Two VCRs
One monitor
Headphones

Put the cable end into the video-out terminal of the player (play machine). Put the other end of the cable into the video-in terminal of the recorder (record machine). Connect the audio-out terminal of the player to the audio-in terminal of the recorder in the same way with another cable. Now plug a cable into the video-out of the recorder and connect the other end to the video-in of the monitor. With this setup, the audio from both play and record machines will be heard through the headphones plugged into the recorder since the monitor has no built-in audio amplifier. See Figure 6–5.

Basic Edit System 2:   Use this setup if you have

Two VCRs
One monitor
Audio amplifier and speakers

This is the same setup as system 1 except we have added an audio amplifier and speakers so that we can hear the sound without the headphones. To set this up, complete the system 1 setup. Then, using another cable, connect the audio out of the recorder to the phono, tape in, or line in (not mic in; that input is for a microphone) of the audio amplifier/speaker system. See Figure 6–6.

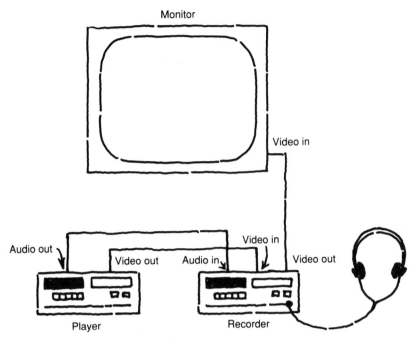

Figure 6-5 Basic edit system 1 includes two VCRs, one monitor, and head-phones wired as shown.

Basic Edit System 3:   Use this setup if you have

Two VCRs
One TV

In another variation on the basic system, you can eliminate the need for an audio amplifier and speakers or headphones by using a TV with a built-in speaker system in place of a monitor. First, connect the video out of the player to the video in of the recorder. Then connect the antenna out (VHF) of the recorder to the antenna in (VHF) of the TV. The RF signal of the TV will combine the audio and video signals so that you will hear the audio through the TV's audio system when you tune the TV to channel 3 or 4. See Figure 6-7.

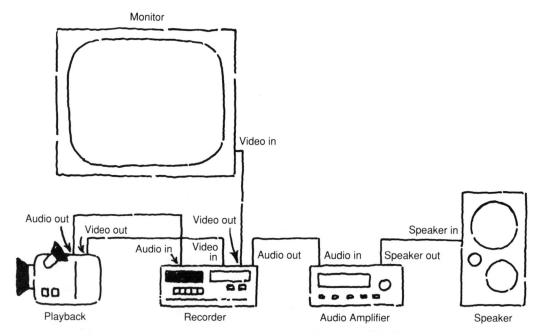

Figure 6-6 Basic edit system 2 includes two VCRs, one monitor, and an audio amplifier with one or more speakers wired as shown. In the drawing, a camcorder serves as one of the VCRs.

Basic Edit System 4: Use this setup if you have

Two VCRs
One TV/monitor combination

In still another variation, you can use a TV/monitor combination to hear the audio. Connect the video out from the player to the video in of the recorder. Then connect the audio out of the player to the audio in of the recorder. Then connect the video out and audio out of the recorder to the video in and the audio in of the TV/monitor combination. See Figure 6-8.

Basic Edit System 5: Use this setup if you have

Two VCRs
One *or* two TV/monitor combinations *or* one or two TVs *or* one TV and one TV/monitor combination

If you want to see the pictures and hear the sounds from both the play machine and the record machine simultaneously, you will need two visual hookups,

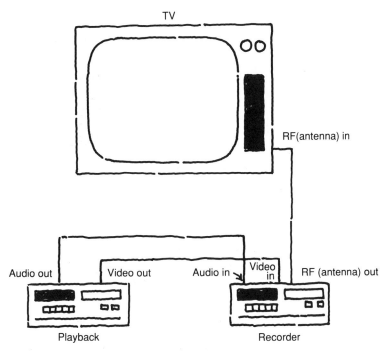

Figure 6-7   Basic edit system 3 includes two VCRs and one TV wired as shown.

either one TV and one monitor (TV/monitor combination) or two TVs or two TV/monitors. With this setup, the TV or monitor wired to the player is designated as preview. The TV or monitor wired to the recorder is designated as program.

If you have one TV and one TV/monitor combination, connect the play machine's antenna out (VHF) to the antenna in (VHF) of the TV. Next connect the video out of the player to the video in of the recorder and the audio out of the player to the audio in of the recorder. Now connect the video out of the recorder to the video in of the monitor and the audio out of the recorder to the audio in of the TV/monitor combination. See Figure 6–9. Any variation on this setup, for example, two TVs, would be connected the same way.

Basic Edit System 6:   Use this setup if you have

Two VCRs
TV/monitor combination with dual video/audio inputs

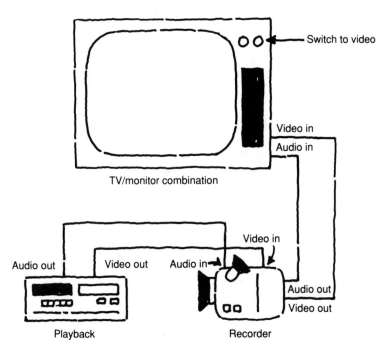

Figure 6-8 Basic edit system 4 includes two VCRs and one monitor/TV combination wired as shown. In this drawing, a camcorder is used as one of the VCRs.

Some TV/monitor combinations have two sets of video/audio inputs that can be accessed individually with a switch or remote control. If you have this type available, connect the player's video out and audio out to video 1 in and audio 1 (or left) in on the TV/monitor. Then connect the recorder's video out and audio out to video 2 in and audio 2 (or right) in on the TV/monitor. This setup will allow you to see either machine as you like by switching from one to the other. See Figure 6-10.

Getting Ready for the Edit

Once you have connected your basic system, test it to see that everything works the way you expect it to *before* you do the edit exercises that follow. Select recorded videotape for the play machine and a work tape for the record machine. A work tape is just what it sounds like, a tape on which to practice edits or to

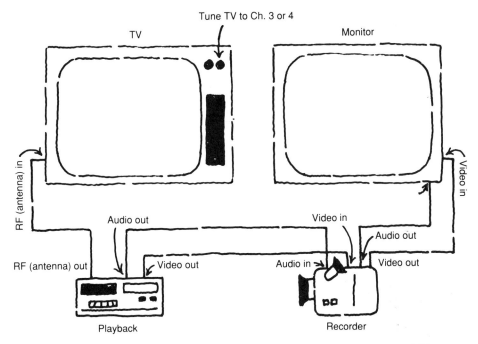

Figure 6-9 Basic edit system 5 includes two VCRs and two of any combination of TVs or TV/monitor combinations, wired as shown. In the drawing, the setup includes one TV and one TV/monitor combination. One of the VCRs is a camcorder.

make a B roll or dub over audio from another source or create titles or whatever. It is a scratch tape for getting ready to do an edit.

Since this is a test, use videotapes that do not have anything of importance recorded on them, just in case material should be erased or recorded over. Check the recorder cassette to be sure that it has the protection tab or button in place so that you can record on it. If it has been removed, put the button back in or put tape over the hole where the tab used to be. See Figure 6–11.

Testing the Setup. Test the system by answering the following questions. The answer to all should be yes. If they are not, check your setup.

1. Are all the units connected to the power outlet and turned on?
2. Press *Play* on the player. Do you see the pictures, (a) on the TV or monitor if you are using basic edit systems 1 to 4 or (b) on the preview TV/monitor if you are using basic edit system 5 or (c) on the TV/Monitor when switched on using basic edit system 6? Do you hear the sounds through the TV/monitor or audio system or headphones?
3. Press *Play* on the player and *Record* on the recorder. Do the pictures and

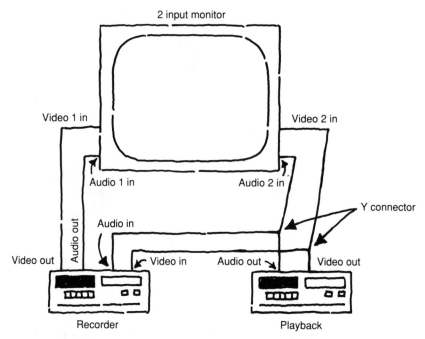

Figure 6-10   Basic edit system 6 includes two VCRs and one TV/monitor combination, with dual video/audio inputs wired as shown.

the audio from the player record on the videotape in the recorder? Do you see these pictures as they record on the TV/monitor?

Learning the Machine's Dynamics.   The next steps require some practice to learn the player's and recorder's dynamics. The typical VCR goes through these steps automatically from the play mode to the record mode:

1. The VCR is in the play mode, moving the videotape from the supply reel to the take-up reel around the play head drum. The tape's velocity is being determined by the control track, and the play head is turned on.

2. You press *Pause* and the tape's forward motion stops and remains around the play head's drum. The play head remains on. When you release *Pause,* the tape plays, picking up where it left off, with the control track control-

Figure 6-11   If the tab has been broken off a cassette, put a piece of tape over the hole so that you can record on the tape again.

ling the tape's velocity at the correct position of the spinning play head. This position is important because the play head turns on to play that frame and then turns off when it is finished and prepares to turn on again to play the next frame. When the head is turned off, this is called the vertical blanking interval.

3. When you press *Record* in the play pause mode, the machine turns off the play electronics and turns on the record circuits, which are the same heads but they are now recording rather than playing. The tape's velocity is still being controlled by the control track if there has been a previous recording, so it moves to the correct position to start recording. On some machines, this process happens in the reverse order; that is, the tape is paused where it is and the positioning happens when the *Record pause* is initiated. This assumes that the VCR is not being controlled by an edit controller, and does not have any built-in edit features, and is a basic VCR.

The basic VCR does not backspace or preroll. It will just back up a few frames so that it has a few control track pulses to get the tape going from the pause mode. An edit from this mode will work most of the time on this type of VCR, but *not* every time you try it.

For the videotape to record properly, it has to get up to speed. To get up to speed, it has to have a running start. To get a running start, it has to back up before it can run forward to start recording. When it starts recording, it covers over part of the last shot that was recorded.

This backspacing will also affect your timing. It means that you will have to make an allowance for how much of the previously recorded shot will get covered by the next shot. In other words, how much more of the first shot needs to be recorded to get to the edit-out point that you want? Use Exhibit 6–1 to check backspacing on your VCR.

Something else you need to know is when the videotape in the recorder and player actually starts to move forward in the play and record modes from the pause mode and how long the VCRs will stay in pause before releasing automatically. This knowledge is vital in order to develop the timing for releasing both machines from their respective pause modes at the right time. This basic setup has no computer control, but rather is strictly punch and crunch. You will need to time your button pushing to the time required for the machines to reach the edit-in (record) point.

Exhibit 6–1 Determining VCR Backspacing

Backspacing occurs when the machine rewinds the tape a specific amount of time from the edit-in point to give the VTR/VCR enough time to get up to speed before recording begins at the edit in. A machine backspaces so it can preroll. To determine how much your record machine backspaces, do the following;

1. Put the edit master into the play machine and use it as your source (original) material. Select a shot, on your edit master that cuts to another shot.
2. Record that shot, including the cut, onto another tape in your record machine. Push *Pause* on the record machine exactly 5 seconds after the cut is recorded.
3. Rewind the videotape in the record machine and cue it up to the exact end of the recorded shot that is, 5 seconds after the cut.
4. Put the machine in record pause.
5. Select another shot on the player and put the player in the play pause mode.
6. Punch *Pause* on both machines. Record for about 30 seconds and punch *Pause* on the recorder and *Stop* on the player.
7. Press *Stop* and rewind to the beginning.
8. Press *Play* and watch the tape. When you see the camera cut, press *Pause.*
9. Press *Play* and determine how much time passes between the cut and the next edit. Deduct this amount of time from your original 5 seconds.

The difference is the amount of time that you have to add to your selected shot length to allow for the overrecording that is a result of the VCR's backspacing.

Each machine will give you different clues as to when the forward movement from pause starts. Use the counter to see in which direction the videotape is moving and what kind of delay there is. On some VCRs the counter will roll backward for all *or* part of a revolution or number, and then revolve or change to the next higher number before rolling forward. Other counters will reverse to the next four or five lower numbers before they advance. On some machines, how the counter reacts depends on the speed at which the tape was originally recorded. In other words, if the tape was recorded at the slowest speed, the counter will move slower than it would if the tape were recorded at a faster speed.

Some other clues that you might want to be aware of are the sounds that the machines make, that is, clicks and whirs. Some sounds will give you different indications in the play or the record mode even if they are the same model.

The point of knowing the mechanical characteristics of the machines is that you will be punching the pause buttons to start the player and the recorder, the punch and crunch method, and to make the final production as good as you can, *your* timing must be as perfect as you can make it. For instance, if your play machine from the play pause mode moves back three counter numbers after you punch pause before it moves forward, and your record machine does the same thing in record pause, you can punch both at the exact same time.

On the other hand, if your play machine rewinds three numbers and the record machine has not rewound at all, you have to punch the play machine slightly before you punch the record machine in order to make a good edit or at least a close edit. Close is about as good as you can do with a punch and crunch system since you are doing everything on the fly.

Be aware that neither machine will stay in the pause mode for more than a few minutes before releasing automatically, going into the stop mode, and possibly rewinding the videotape into the cassette. Manufacturers have designed this into the equipment so that the videotape does not stay over the heads for too long. This automatic pause release function can play havoc with your edit since, if it happens, you will have to cue up both machines again before proceeding with the edit. What is the answer? Be ready and be quick. That is the only way to do punch and crunch editing.

Most VCRs do not start to record cleanly from the record pause mode on *blank* videotape because no control track is recorded on the tape. Control track is a series of electronic pulses, one for every frame, recorded on the edge of the videotape along with the video/audio signals. Control track tells the tape transport at what velocity to play back the tape. If it is not at the precise speed, you will see a band of snow in the picture, a glitch at the spot that the control track lost control of the videotape.

Ideally, the control track should be continuous from one end of the edit master to the other for it to play back without problems. This is difficult, at best, using the punch and crunch system since it is an assemble edit system and thus records new control track with every new video edit. Everytime the recorder is paused, it stops recording the new control track. If your VCR is not designed for editing, when the machine goes into the record mode for the next edit, it might not record the new control track in the exact place that it should, causing a glitch. See Figure 6–12.

The way to overcome this problem is to always record more video than you have selected for that shot; then press *Pause* (never *Stop*) on the recorder. After the VCR has paused, press *Stop;* then play and rewind the tape to the place that you want the edit-out point and pause; then record. All assemble edits like this will record control track at the edit in but leave the control track ragged at the end of the recording.

Camcorders and some VCRs are designed to rewind a few frames when they are put in the record pause mode. In that way, they have the previously recorded control track to get the videotape to the correct speed before they start to record the new video and control track. Others just pause the tape and delay rewind until they are released from pause.

All this means that to get rid of an edit glitch, the videotape must have video recorded on it prior to the edit-in point or a black and white ragged band may appear where the new recording begins. The solution then is to overrecord every shot so that the next edit records over a part of the previous one. Or you can black a tape. See Exhibit 6–2.

Another problem you may face with punch and crunch editing is the rainbow effect. This is a rainbow moiré that appears in the recorded video at the edit-in point and continues for a few frames. See Figure 6–13. This moiré is caused when the previous video has not been completely erased before the record head begins recording the next shot. This leftover video appears as a narrow vertical line divided into red, green, and blue bands recorded on the edit master. There is no way to eliminate this problem completely except by using a VCR with a flying erase head.

A VCR with a longitudinal erase head will start to erase the helical scan

Figure 6–12 A VCR not designed for editing may not record new control track at exactly the point where the existing control track ends, causing a glitch on the incoming video.

Figure 6-13 A rainbow moiré may record over the picture at the edit-in point on a system not designed for editing.

video information from the bottom of the scans to their top as the tape is being transported horizontally past it, leaving part of the scan undisturbed. The erase pattern continues progressively toward the top of the next scan, and then the next, and so on until it finally erases a complete scan at some point, depend-

Exhibit 6-2 Blacking a Tape

To record video black, you cannot just put the machine into the record mode. You must feed it a video signal. You can get video black from an SEG or any camera or from another blacked tape. The process is similar for all three. For our example, we will use a camera/camcorder.

To black an edit master, connect your camera or camcorder video out to the record machine's video in. Leave the black lens cap on the camera/camcorder. Turn the camera/camcorder on.

Rewind the edit master in the recorder to the beginning. Press *Record* on the recorder and do not stop recording until the entire tape has video black recorded on it, front to back.

Use an ac connector to power a camera/camcorder to ensure that the process will not run down the battery and shut down during blacking the tape. *Note:* Always use the same machine to record the BEM (Blacked Edit Master) that you will use as the record machine in your edit system. This will eliminate any problems from machine to machine.

If you are using a camera/camcorder to record the BEM, be aware that the mic will automatically turn on during the recording process. Disconnect the mic input so that you do not have unwanted audio recorded on the BEM. If you cannot disconnect the mic input, plug a cable adapter into the mic input of the camcorder. This will turn off the camcorder's mic. See Figure 6-14.

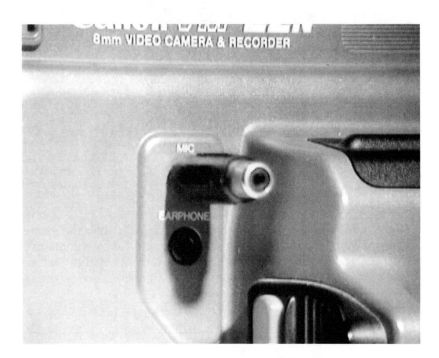

Figure 6-14 To black a tape on a camcorder, disconnect the mike if possible or there will be audio recorded as the video black is recorded. If you cannot, plug a cable adapter into the MIC INPUT to turn off the built-in microphone.

ing on the design of the VCR, and then continues to completely erase the whole tape.

When the new video starts to record, it traces a new scan in the former partly erased ones and progresses until it finally reaches the ones that have been completely erased. As you will see when you play the tape, the moiré is caused by part of the previous scan and the current one being played together.

STEPS TO SIMPLE VIDEO EDITING

To learn more about the basic editing techniques and how the basic equipment works, practice editing using the following step by step process of *simple* video editing using the basic edit system. Prepare a brief edit decision list (EDL) from which to work. Choose 10 different shots for the EDL. Once this is complete, begin.

An Edit Exercise

1. Turn on all the equipment.

In the Play Machine

2. Protect the original material from accidental erasure by removing the tab or button on the recorded tape. Three-quarter-inch cassettes and some Beta cassettes have buttons; all other cassettes have tabs that break off. Reel to reel tapes have no protection.
3. Load the original material tape with the first shot into the player. Play a few minutes of this tape to check the tracking. This is especially important if you are using a different machine to play it on than it was originally recorded on. Adjust the tracking if necessary.
4. Rewind the original material to the beginning of the tape.
5. Set the player machine's counter to zero.
6. Set the *edit switch* on the player (if it has one) to the correct position.

In the Record Machine

7. Check the videotape you will be using as your edit master to be sure that the tab or button is in position. If it has a button and it has been removed, put it back in. If it has a tab and it has been removed, cover the hole with a piece of tape. The machine will not record on the tape when these have been removed. Load the videotape into recorder.
8. Fast forward the videotape for approximately 30 seconds; then rewind it to the beginning. Zero the counter.
9. Set the record machine to the SP speed (fastest) to give you the best possible picture.
10. Record black on the videotape in the recorder for about 30 seconds. Then press *Pause.* Do not ever start your editing at the exact beginning of the videotape. The machine will not have enough preroll time to get the tape up to playing speed. Thus the machine will not record properly. The recorder should now be in the record pause mode.

The edit system is now set up to assemble edit the selected shots from the original material as noted on your EDL. All transitions between shots will be cuts only because that is all the system is capable of doing.

Editing Steps Continued

11. Put the player in the play mode or fast forward until you reach the first shot. Punch *Pause.* Check to see that the counter number corresponds to the number on the EDL and visually looks like the right shot.

12. Check the recorder to be sure that it is still in the record pause mode. Reset the counter to zero. By zeroing the counter at the first edit, the machine will keep a running timing of the program for you, if your equipment has control track readout. This is not essential, but it is helpful as the edit proceeds to know how long the program is with each next edit.

13. Punch *Pause* on *both* machines. The play machine will begin playing and the record machine will begin recording. Depending on the dynamics of the machines, you may want to release one of the machines before the other, for example, if one backspaces and one does not.

14. Watch the counter of the player or the video itself, and when it reaches the selected edit-out point, punch *Pause* on the recorder. Punch *Stop* on the player.

15. Based on your EDL, find the next shot on the play machine and repeat steps 12 through 14 (except for zeroing the record machine's counter) until the editing of your production is completed.

That's it—a simple assemble edit using the basic punch and crunch, step by step process of editing. Admittedly, this basic process does not lend itself to 100% accurate editing, but it does cover all the basic steps and some of the basic knowledge required to refine your editing skills and your knowledge of how all the elements work together. The bottom line is: The machines, TVs/monitors, audio system, videotape, and *you* are all interrelated and must function together.

Most professionally produced TV programs start with a fade in from video black. Fades are not possible with the basic edit system, but you can still have a fade up from black in a program assembled with a basic punch and crunch edit system with planning. For example, the VCRs in the system cannot do a fade in, but your camcorder most likely can. Most camcorders have a fade in and fade out feature that can be initiated at the time of the original recording. Shoot your opening shot as a fade in on the camcorder and the fade in will be built in when you go to the edit. This is just one way of using your head to do something that your edit system might not be capable of doing.

Video editing is really an art, not just mechanics. The mechanics, however, can get in the way of the art if you do not understand the relationships. By eliminating some of the potential problems of the mechanics, you can make the art more effective.

INCREASING EDIT ACCURACY

Punch and crunch editing accuracy is a function of your skill, dexterity, and, most of all, timing. Your eyes *look at the pictures* on the screen and the information is instantaneously transmitted to your brain, where you make the decision of where you want to start and stop the edit. You brain tells your finger to punch the button, which it does, and the edit is made or the effect is recorded. Then you rewind the recorder to look at what has been recorded and it is not what you saw at all. What happened?

What happened is that you were too slow in reacting to what your eyes saw on the screen. It took too long to make the decision. It took too long to tell your finger to move. It took to long for it to move and it took too long for the machine to do what your finger told it to do. The entire sequence just took too long to accomplish what you wanted. The total elapsed time from decision to finger punching was maybe 2 seconds. In this particular case, that was too long.

In that 2 seconds, 60 frames of video rolled by, and by the time the actual edit was performed, the shot had moved or the person's expression had changed or something else had happened to throw the timing off so that the edit or the effect just did not work.

In video editing, 60 frames is not even close, and even if the time from the decision to the punch time is cut in half, 30 frames is not much better. The goal is to get as close to the exact frame as you can. In short, to be accurate, you have to be fast *and* prepared.

Here are a few things that can help.

Anticipating the Edit

First, *stop looking at the picture during the edit.* You have already selected the picture that is going into the program; the real problem is that the picture changes every $\frac{1}{30}$ second (or $\frac{1}{25}$ second if you are using SECAM or PAL). How do you get it to start and stop where you want it to?

The answer is that you have to be able to predict when that exact frame is going to be passing by the play head and grab it. To be able to make an accurate prediction, you need information. After you have the necessary information, the only thing left to do is the mechanics to get it there and a little luck. A part of the mechanics is building more time into the amount of time that it actually takes for the process to happen. Call it anticipation time. It is just like anticipating the start of a race: *Get Ready . . . Get Set . . . Go!*

With punch and crunch editing, the control of the edit points lies mainly with the player and its counter. If you have a choice of two machines with different features, you will want to choose the one that allows you to position and pause the original material in the exact edit-in position.

For example, if one machine has a single frame advance and rewind feature, that is the one to use as the player. If you do not have these particular features, you will have to work out the method that works on the equipment that you have. On some machines this might only be three buttons, the pause, play, and rewind, but it can be done. The process just takes a little longer and a lot more patience. The important part to keep in mind is the principle of building in the needed time for you and the equipment to react so that the exact frame or one close to it gets recorded on the edit master.

Follow this simple "Get ready, Get set, Go" plan for maximum accuracy:

Get Ready.   In video editing the *get ready* is knowing the location of the shot on the original material. For our example, the player's counter is the type that reads in hours, minutes, and seconds. The EDL indicates that the shot is located at 0:12:? (0 hours, 12 minutes, and unknown seconds). You roll the edit master until the counter indicates 0:12:00 and punch *Pause*.

Get Set.   With the player in the play pause mode, move the videotape, frame by frame, if you can, to the exact frame where you want the shot to start. Next, look at the player's counter number and write it down on the EDL. You now have an *almost* exact edit-in point for this shot. For our example, let's say that the counter reads 0:12:39 (0 hours, 12 minutes, 39 seconds). The counter is now within the range of 30 frames of the shot. You know this even though the counter system does not show you frames. Remember that 30 frames roll by in 1 second (25 frames on PAL or SECAM). If you do not have a frame-advance switch, this will have to be close enough since it is as close as you are going to get to the exact frame.

If you have a frame-advance switch, you can get to the exact frame in point by taking one more step. Jog the videotape forward a frame at a time. You count the number of frames by how many times you punch the frame-advance button to make the counter turn to the next highest number. One punch equals one frame. Let's say that you punch and count 15 frames to change the counter from 0:12:39 to 0:12:40 (remember, the last two numbers are seconds). Now you know that the frame that you want to edit in is halfway between 39 seconds and 40 seconds on the player's counter. There are 30 frames to 1 second, so halfway is 15 frames. The exact NTSC frame location on the original material is 0:12:39:15 (0 hours, 12 minutes, 39 seconds and 15 frames).

The next step in "get set" is to know as accurately as you can when the recorder actually starts to record after you release it from the record pause mode. For our example, let's say that it takes exactly 2 seconds from the time that you release *Pause* for the recording process to start.

Rewind the player about 30 seconds or more if you prefer more time to anticipate the counter advancing to the frame edit-in location. Let's say 30 seconds is enough anticipation time and the counter reads 0:12:09. That is the edit-in time of 0:12:39: minus 30 seconds. Put the player in the play pause mode.

Again, if you have a frame-advance switch, jog the player until it reads exactly 0:12:09 for our example. So, you rewind a frame at a time until the counter changes to 0:12:08 and then advance to the next frame that changes the counter to 0:12:09. The player is now set to give you exactly 30 seconds and 15 frames of preroll time before the exact frame plays. You know this because you are parked at 00:12:09:00 and your edit-in point is 00:12:39:15. Subtract 00:12:09:00 from 00:12:39:15 and the answer is 30:15 = 30 seconds and 15 frames.

You have determined that the recorder has a 2-second preroll, so this amount of time is deducted from the player's preroll time; thus your total time to anticipate the incoming edit is 28 seconds and 15 frames. The 28 seconds is an exact time or as exact as you and the player can make it. The 15 frames is your intuition, judgment, guess work, and luck.

Go. Press the player's pause button to release it from play pause. *Don't watch the picture.* The timing is too critical. Concentrate on the player's counter. When it reaches 0:12:37 plus your time judgment about the 15 frames, punch the pause button on the recorder. How soon to punch pause on the recorder *after* you see 0:12:37 on the player's counter is a skill that you develop with practice. The odds of catching the exact frame are not in your favor, and the goal of being within a few frames is attainable with the right equipment.

The Finish Line. This is the edit-out point. You have carefully selected and cued the player to the edit-in point, but it is just as important to do the same for the edit out. It is done with the same steps as the edit in, except you are locating the exact frame for the edit to end.

The anticipation time to punch *Pause* on the recorder is finite. It is only the length of the shot plus the amount of time that you need to overrecord to compensate for the frames that will be covered by the next shot. For example, let's say that your shot from edit in to edit out is 14 seconds and 10 frames. That is all the anticipation time you have before pressing *Pause* on the recorder. It is not much, so *get ready, get set, and go!*

Again, *don't watch the picture.* Watch the counter on the player. When it reaches 0:12:53 (12 minutes and 53 seconds) plus your feel for 10 frames and your overrecord allowance, punch *Pause*. That is, you edit-in point of 0:12:39 plus the shot length of 0:00:14 added together gives you the counter total for the edit-out point of 0:12:53. The 15 frames for the edit in have been compensated for at the point that you pressed *Pause* on the recorder to start the edit. The 10 frames on the edit out must be taken into consideration when you punch *Pause* to stop the recording process.

Step by Step to Edit Accuracy

Step by step, here's how to increase your editing accuracy:

1. Locate the selected shot as indicated by the counter number on your EDL.
2. Load the original material into the player and rewind to the beginning.

3. Reset the player's counter to zero.

4. Play or fast forward to the counter number indicated on your EDL. Press *Pause.*

5. Cue the original material up to the edit-in point. Write the counter number on your EDL.

6. Jog the original material a frame at a time to the next highest counter number. Count the number of frames as you advance the original material. Add this number to the notation of the exact shot location on your EDL.

7. Determine the length of the shot using the same method used to determine the exact edit-in point. Add to this your allowance.

8. Write the shot length on your EDL.

9. Set your record machine to record at the SP speed.

10. Put the recorder in record pause to prepare it for the edit. *Note:* If you can do steps 1 through 9 before the recorder automatically releases itself from the record pause mode to the stop mode, set it in record pause as the first step.

11. Rewind the original material the amount of time that you need for anticipation time.

12. Jog the player back to the next lower number.

13. Jog the player forward to the exact next highest number.

14. Calculate the exact amount of anticipation time by deducting the amount of time that it takes for the recorder to start recording after you punch *Pause.*

15. Release *Pause* on the player and watch the counter to indicate the next step. This is the number that you calculated in step 14.

16. When the calculated number reads out on the counter, release record pause on the recorder. If you have an effect such as a fade in or wipe in, it should start at the end of the recorder's preroll time. Add the time for the effect to the exact edit-in point. If you have a fade out or wipe out, it starts before the counter reaches the edit-out point. Add the time for the effect to the shot length. If the shot fades in *and* fades out or wipes in *and* wipes out, add the effect-in time to the exact edit-in point and the effect-out time to the shot length.

17. Watch the player's counter, and when it advances to indicate the shot length, press *Pause* on the recorder.

18. Press *Stop* on the player.

19. Repeat steps 1 through 18 to continue to the next edit.

Zero Counter Method

The most refined way to increase editing accuracy is to use the zero counter method; that is, you reset the counter to zero after you located the exact frame for the edit-in point. With this method, you can also use a counter that reads in arbitrary numbers, as well as the real-time counter that we used in our example.

All the steps are the same in this method as they were for our example, including locating the edit-out point, *except* that when you have the edit-in frame in play pause, push the counter reset. If we use our shot location from our first example, the counter will change from 0:12:39 to 0:00:00. The advantage with this method is that it accounts for the 15 frames that the counter does not show you, and you do not have to guess when to release the recorder. The counter now reads 0:00:00 and the machine is in play pause at the exact frame that you want to grab. If your counter is arbitrary, it will read 0000.

When you rewind the player for the 30-second anticipation time, the real-time counter will read -0:00:30 (*minus* 0 hours, 0 minutes, and 30 seconds). Taking into account the 2- to 5-second preroll that the recorder requires to start the recording process, punch *Pause* when the player's counter reads -0:00:2 (*minus* 0 hours, 0 minutes, and 2 seconds). The recording should start when the player's counter reads 0:00:00 and near or at the exact frame that you selected.

If your counter is arbitrary, then you must use the factor that you calculated for the number changes that we described previously. If your original material was recorded at a speed other than SP (fastest), this will affect the factor.

When using the zero counter method, it is important to keep track of the amount of time that is reset off the counter so that you can find the next shots on your EDL. Let's say the next shot is located at 0:26:?. You need to deduct the 0:12:00 that you rolled off from the previous shot to be able to locate the next shot easily. The next shot would now be located at 0:14:00. That is, 12 minutes subtracted from 26 minutes equals 14 minutes.

Follow these steps when using the zero counter method:

1. Locate the selected shot as indicated by the counter number on the EDL.
2. Load the original material into the player and rewind it to the beginning.
3. Reset the player's counter to zero.
4. Play or fast forward to the counter number indicated on the EDL. Press *Pause.*
5. Move the original material to the exact frame at which you want the edit-in point. Write the counter number on the EDL.
6. Determine the length of the shot using the same method used to determine the exact edit-in point. Write the shot length on the EDL.
7. Put the recorder in record pause.
8. Find the exact frame for the edit-in point and pause the player.
9. Reset the play machine's counter to ZERO.

10. Rewind the original material the amount of time that you need for anticipation time.

11. Calculate the exact amount of anticipation time by deducting the amount of time that it takes for the recorder to start recording after you punch *Pause.*

12. Release *Pause* on the player and watch the counter.

13. When the calculated number reads out on the player's counter, release the record pause on the recorder.

14. Watch the player's counter, and when it advances to indicate the shot length, press *Pause* on the recorder.

15. Press *Stop* on the player.

CONCLUSION

The configuration of editing systems varies from a simple two-machine punch and crunch, cuts-only system to a multitape source, digital effect system. The editing possibilities of any system are limited by the equipment's capabilities and the editor's knowledge and expertise at utilizing those capabilities.

The most basic of editing systems consists of two machines: a player and a recorder. It also includes either one or two TVs or monitors and an audio speaker system if one is not a part of the TV/monitor hookup. It is a cuts-only system capable of doing assemble edits only.

The biggest problem with the basic editing system is its inaccuracy. There is no computer control, so the timing of an edit is strictly dependent on the editor's ability to punch the button at the right time. Another problem is machine backspacing. Without a computer controlling this, it is up to the editor, again, to compensate for it and slide the edits one way or the other to make sure all the material that he or she wants recorded gets recorded *and* stays recorded. It is entirely possible that a recorded edit will get clipped by the next edit because of backspacing, with some of the wanted recorded video getting erased.

Basic editing also relies on the pause buttons on the record and play machines. To get ready for the edit, both machines are placed in pause. If the editor should take too long cueing up a machine, the pause on the other machine may be released. All machines have built-in automatic pause release so that videotape will not stay over the record/play heads for too long. If this happens, the editor has to start over, recueing that machine.

If the record VCR does not have a flying erase head, every edit will begin with a rainbow moiré. This glitch is a combination of the old signal already recorded on the tape and the new signal coming in.

Editing is an art, even when it is basic. Like any art, the mechanics must be learned before you can perfect the technique of doing it.

TO DO

What have you learned about basic editing? If you have not already done so, test yourself by doing the following:

1. Hook up a basic edit system using the equipment that is available to you.

2. How does the system perform? How much backspacing does it do when you release pause?

3. Black a tape and make several edits on it. Do the same with a blank tape that is not blacked. Compare the edits made on both. Which works better with your edit system?

7

Advanced Editing Systems

Beyond the punch and crunch editing system is a whole world of editing possibilities. There are VCR/VTRs with built-in editing features, edit controllers, SEGs (special-effects generators), CGs (character generators), computer animation, and all kinds of black boxes. All have only one purpose—to make your program better both technically and creatively.

The more of these bells and whistles that you have, the more sophisticated your editing system will be, with the quality of the program rising proportionally. Not many of us will ever have everything, but we can all build on the basic setup, as our budgets and time allow. The first thing you might do is add a SEG.

THE NEXT STEP: SEGS

SEGs are designed basically to perform special effects, to color correct the picture, and to fade in or out the video or audio. They come in all different sizes, from the home consumer version, shown in Figure 7–1, to the high-tech ones offering sophisticated digital effects, shown in Figure 7–2. Some include a built-in digital frame synchronizer that allows effective mixing of video sources, such as VCR/VTRs and cameras, with no need for time base correction. This may include the mixing of composite and Y/C signals with Y/C separation and the mixing of variable picture resolutions. The features that you might find on an SEG are discussed next.

Figure 7-1 Consumer SEGs are becoming more and more sophisticated. The one shown, the Panasonic WJ-AVE5, can do over 100 special effects including dissolves, superimposed images, picture-in-a-picture, 98 wipe patterns, and more. (Photo courtesy of Panasonic Audio/Video Systems Group.)

Figure 7-2 The Ampex ADO 100 digital effects system pictured here provides smaller broadcast, post-production, and corporate industrial facilities with traditional ADO picture quality, extensive effects. In the picture, the ADO is putting a box insert into a live picture. (Photo courtesy of Ampex Corporation.)

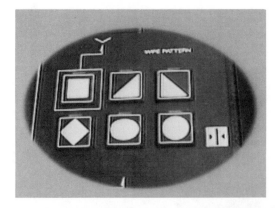

Figure 7-3 Wipe patterns available on SEGs can be simple, as shown, or very complex, with as many as 99 different wipe variations.

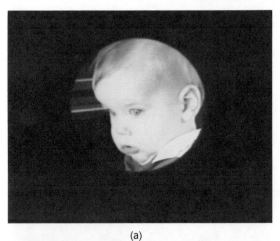

(a)

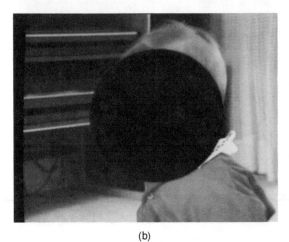

(b)

Figure 7-4 Wipes can generally be brought in either from the edge of the screen, as seen in (a), or from the center of the screen, as seen in (b).

Wipe Patterns

The most popular option on a special-effects generator seems to be wipe patterns. These wipes come in various shapes, as shown in Figure 7–3. In basic SEGs, you can wipe from a picture to a variety of solid colors, or vice versa, for example, from a picture to the color blue or from the color blue to a picture. More sophisticated SEGs can wipe in various patterns from one video source to another, for example, from one picture to another.

Wipe patterns are drawn on switch buttons, which when depressed set up to do that type of wipe. A fader bar is then used to bring the wipe in or out. Wipes can generally be brought in either from the center of the screen or from the edges, as shown in Figure 7–4. For example, if the wipe is a circular pattern, the picture is brought in in a circle either from the outside or from the center. Also, some controllers offer the editor the option of a hard-edged wipe, shown in Figure 7–5, or a soft-edged wipe, shown in Figure 7–6.

Negative/Positive Image Reversal

Color pictures can be changed to a negative image, similar to a photo negative. See Figure 7–7.

Digital Strobe

The digital strobe capability creates stop action or slow-motion-type effects by freezing frames of video. The sampling rate can be adjusted to the desired speed. See Figure 7–8.

Figure 7–5 Wipes can have a hard edge.

Figure 7-6 Wipes can have a soft edge.

Freeze Frame

The freeze-frame function digitally stores a frame of video to provide clear still images without interfield flicking. This feature allows you to insert video photos into your productions.

Superimpose Images

Some SEGs allow you to superimpose a title over a picture, as shown in Figure 7-9, or to key the picture into the title, as shown in Figure 7-10.

Monochrome Effect

The color level is reduced, changing the picture into a monochrome picture. This is sometimes called painting or matting. See Figure 7-11.

Monotone Effect

A color can be added to the monochrome picture, for example, to make it intensely blue or red or green.

Mosaic Effect

The picture information is digitally processed to produce a mosaic or pixelized effect. In other words, the picture is broken down into different sizes of squares, with the effect increasing in intensity until the picture is unrecognizable. See Figure 7-12.

(a)

(b)

Figure 7-7 Some SEGs have the ability to reverse the picture from positive to negative.

Fade to or from Black or White

When this fade is used, both the video and audio fade in or out together. The timing of this fade in or out is usually preset, for example, a 3-second fade in. See Figure 7-13.

Video and Audio Fade in and out

Most SEGs will allow the operator to fade the video in or out either manually with a fader bar or with a variable or preset timer that is activated by a switch. Some will fade the audio separately and up or down to match the video.

Figure 7-8   A digital strobe effect is created by shifting the image back and forth rapidly.

Some SEGs can time the video signal so that wipes and other video switching can be done glitch-free, while others are designed to process asynchronous video without which the edited master would have a glitch everytime the video source was switched from one to another. If the unit is capable of performing a wipe from the original material to a video color generated by the unit and then back to another shot, each transition would have a glitch if the video signals were not in sync with each other or if there were no edit at the switch points.

Figure 7-9   Superimpose titles over a picture as shown.

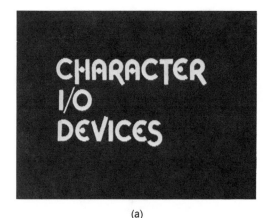

(a)

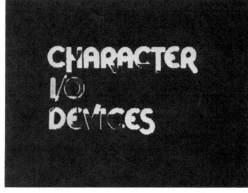

(b)

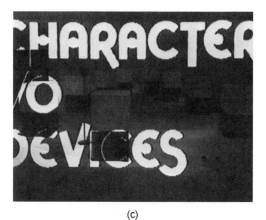

(c)

Figure 7-10 Some SEGs can key the picture into the title. In (a), the title is solid. In (b), the video is seen through the title letters. In (c), the video is still seen through the letters *and* a dissolve to the picture is in progress while the camera zooms in on the title.

Video Enhancement

Most units have a video enhancement circuit. The amount of change in the look of the picture is controlled by a switch that varies the amount of signal that is processed. It can be a fader, a joystick, or a simple knob that the operator turns. Some units will make the picture turn black, at one extreme, to washed out at the other.

Color Correction

Some units include circuits to manipulate the red, blue, and green hue and saturation of the chroma of the video signal. The amount of correction possible depends on the unit's capability. The unit may split the screen during the correction process, with the half of the picture on one side showing the original picture

(a)

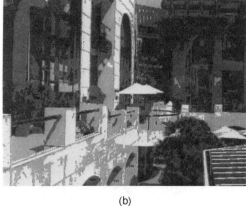

(b)

Figure 7-11 When color levels are reduced, the picture is given a matted or painted effect.

(a)

(b)

(c)

Figure 7-12 The picture information is digitally processed to produce a mosaic or pixelized effect. The effect can continue breaking the picture down until it is unrecognizable, as shown in this series of photographs.

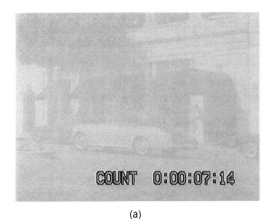

(a)

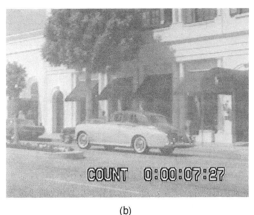

(b)

(c)

Figure 7-13   A fade in or out from black or white fades in or out both video and audio. Note that the fade in from white takes 43 frames; that is, (a) is at 0:00:07:14 and (c) is at 0:00:08:27; thus the effect lasts 01:13, or 1 second, 13 frames, or 43 frames.

and the half on the other side showing the picture as corrected, as shown in Figure 7–14. This allows you to see the difference between the two.

Digital Video Effects

High-tech editing suites may include special-effects generators that transform the analog signal into digital so that it can then perform elaborate effects, such as flips, spins, twists, compressions, expansions, distortions, rotations, and just about anything else you can think of. Some of these effects are illustrated in Figure 7–15.

An Editing Exercise

Before we go on, let's take a moment out to see how an SEG adds to the basic editing system. An SEG can be added to the basic editing system by wiring the

Figure 7-14 Some units split the screen, with one half showing the original picture and the other half showing the color corrected picture.

(a)

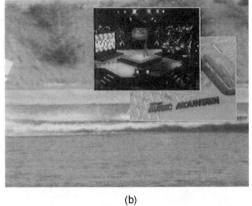

(b)

(c)

(d)

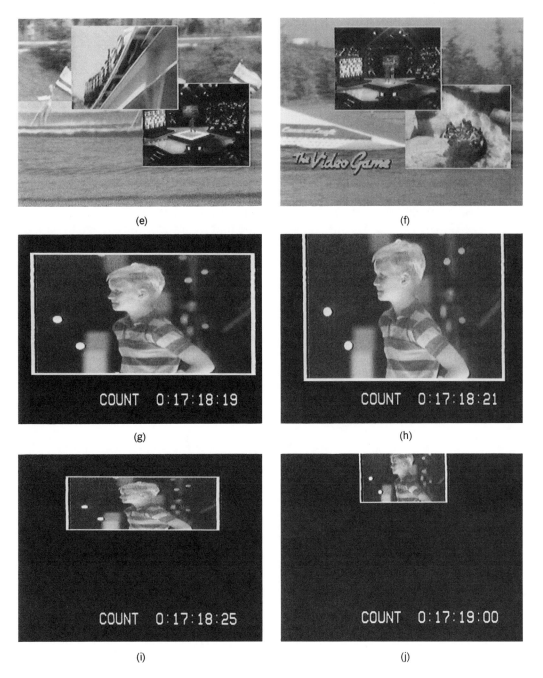

(e)

(f)

COUNT 0:17:18:19

(g)

COUNT 0:17:18:21

(h)

COUNT 0:17:18:25

(i)

COUNT 0:17:19:00

(j)

Figure 7-15 This series of photos illustrates digital video effects (DVE). In the first series, (a) through (f), two boxes are superimposed over the base picture. Other moving video is put into these boxes. The boxes are then moved and shuffled, increasing and decreasing in size. A graphic, "The Video Game," is added. The final effect can be seen in Figure 1–4. In the second effect, (g) through (j), the picture is squeezed over black horizontally, then squeezed vertically, then horizontally, and finally vertically as it decreases in size. Note from the time code on the pictures that the series of photos begins at 0:17:18:19 and ends at 0:17:19:00, lasting :11 or 11 frames.

video out of the play machine to the video in of the SEG. Connect the video out of the SEG to the video in of the record machine. If the SEG has audio capabilities, connect the audio out from the play machine to the audio in of the SEG and the audio out of the SEG to the audio in of the record machine. See Figure 7–16.

SEGs vary in sophistication. For this example, look at how the unit can perform video fades. By processing the video signal from the play machine through the electronics of the SEG, it can be changed from video pictures to video black. On some SEGs the change from or to video black is done with a variable switch called a fader or fader bar. As the fader slides from one extreme to the other, the picture becomes darker until it is completely black. When the fader is moved in the other direction, the picture is revealed as the black diminishes. See Figure 7–17. The fade rate or the transition time that it takes for the change to happen is determined by how fast the fader is moved by the operator.

On some SEGs the fade rate can be preset to have a particular transition rate, for example, a 10-frame fade in. The operator sets the desired rate and then pushes another switch to start the fade-in or fade-out sequence. The SEG does the rest. Some SEGs can only do a fade at a transition rate built into the unit. The editor has no choice about how long it lasts.

When using a fade in, keep in mind that during the transition time the video and audio are not up full but rather are fading in. Thus the transition time should be *added* to the total shot length. For example, if the first shot (first video) is 30 seconds long and you want a 2-second fade up from black, you need to extend

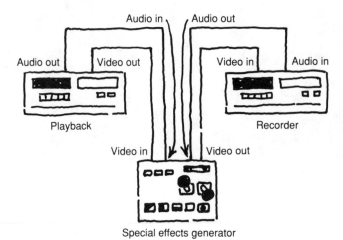

Figure 7-16 A special-effects generator can be added to a basic editing setup as shown.

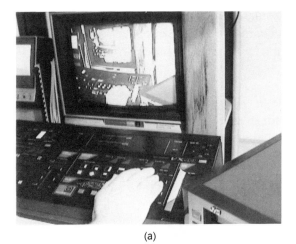

(a)

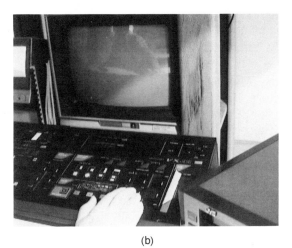

(b)

Figure 7-17 When the fader bar is moved one way, the picture fades to black. Moved in the reverse direction, black fades to the picture.

the shot 2 seconds, making the total 32 seconds. In this way, you will not clip or cut off any action occurring at the beginning of the shot. Fade-up rates vary but are usually no less than 10 frames and no more than 3 seconds.

Using a SEG in your edit system requires that you increase your skill and timing, because you are adding a video effect that must be timed exactly to create the look that you want in the final production.

Try your hand at doing a fade in using the following basic steps:

1. Check the tape in the record machine to be sure that the tab or button has not been removed.
2. Load the tape into the record machine.

3. Fast forward for approximately 30 seconds; then rewind to the beginning.

4. Set the record machine to record at the SP speed.

5. Record black on the tape for about 30 seconds. Press *Stop.*

6. Pull the SEG's fader bar so that you can see the original material pictures on your monitor.

7. Back time 2 seconds from your selected edit-in point on the original material. If you counter reads in real time, subtract 2 seconds from the edit-in counter time. If your counter reads in arbitrary numbers, determine how these numbers convert to real time if you do not already know. (See Exhibit 5.1)

8. Zero the play machine's counter. Now back time an additional 30 seconds. Press *Pause.*

9. Put the record machine in play pause and punch *Record.* The machine should now be in record pause.

10. Reset the counter on the record machine to zero.

11. Move the fader on the SEG to the video black position.

12. Punch *Pause* to release both of the machines from their respective pause modes.

13. Watch the counter on the play machine, and when it reaches zero, move the fader to reveal the picture. Remember that this is a 2-second fade up, so timing is important. If your SEG does not have a preset, it is up to you to get as close as you can to 2 seconds.

14. Watch the counter of the play machine, and when it reaches the selected edit-out point, punch *Pause* on the record machine. Punch *Stop* or *Pause* on the play machine.

These edit steps are basic for all the effects that most SEGs are designed to do.

ADDING AN EDIT CONTROLLER

Edit controllers for VCRs are manufactured by many different companies. Some of these are made by major manufacturers of VCRs and camcorders. Others are made by companies independent of the majors. The thing to keep in mind is that the play and record machines must be compatible with the edit controller or they just will not work together.

Edit controllers are designed to fine-tune an edit with electronic editing capabilities. Most come equipped with a comprehensive selection of automatic computer-controlled editing functions, including the ability to do insert as well as assemble edits and to color process the picture or add graphics. Some even have some built-in effects. All use some form of time code editing, whether it is control track or recorded time code. All this adds one major requirement to your production: *accuracy.*

Adding an edit controller is as simple as purchasing a VCR with one already built in, illustrated in Figure 7–18, or purchasing a separate unit, illustrated in Figure 7–19, compatible with the VCRs you already have. Some of the features you might find on an edit controller are discussed next.

Assemble and Insert Edits

There are two basic modes of editing: assemble and insert. Of the two, assemble editing is the most used, simply because it is the easiest to do. Even a basic editing setup with no edit controller can do an assemble edit. An assemble edit records video, audio, and control track at the same time, in sequence.

An insert edit, on the other hand, allows you to record video only or audio only or both, leaving the already recorded control track in place. *And* the audio can be recorded on one of two available tracks or on both of these two tracks. Also, an insert edit, as its name indicates, can be inserted in between existing recorded material whether it is video or audio. For example, you can edit together a video portion with on-camera audio recorded on the left audio track, usually designated as *L audio* or A1. Then you can go back and put music under the entire segment on audio track number 2, usually designated as *R audio* or A2, disturbing nothing you have already recorded. The types of insert edits you can perform include V + LR (video and both left- and right-channel audio), V (video only), V + L (video and left audio channel), V + R (video and right audio

Figure 7–18 Some high-end VCRs, like the one shown, have a built-in edit controller. The unit shown can do frame-accurate, control track insert or assemble edits.

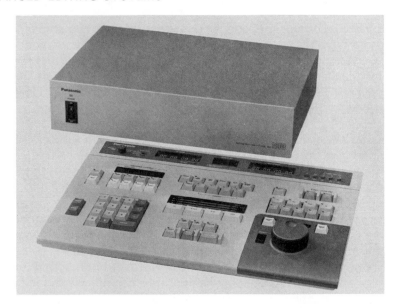

Figure 7-19 An edit controller can be a separate unit designed to control VCRs or VTRs wired to it. (Photo courtesy of Panasonic Audio/Video Systems.)

channel), L (left-channel audio only), R (right-channel audio only), and LR (insert audio on both left and right audio channels).

Some edit controllers have the ability to do automatic insert and assemble edits. Simply, this means the controller will store the edit-in and edit-out points of several edits and then perform them all at once. This feature can save time, but it can also be tricky if you are unsure of your edits, especially with assemble edits.

The ability to do an insert edit adds immensely to a production by giving you the ability to insert titles or graphics, add narration or music, and insert shots that will improve the overall production. An edit controller with the ability to do assemble *and* insert editing is a valuable addition to an edit system.

Marking an Edit

All edit controllers have some way to mark an edit in and an edit out. The mark may be placed electronically on the frame or it may be stored by control track number or time code. When the edit is performed, it begins recording at the edit-in point and, if it is an insert edit, it ends at the edit-out point. If it is an assemble edit, the recording usually ends several frames after the edit-out point, with the record machine immediately recueing to the exact edit-out point. The excess material recorded beyond the edit out in an assemble edit is there to provide control track for the beginning of the next edit.

Automatic Backspacing

When you are ready to perform the edit, you put the play machine and the record machine into pause and press *Edit*. The record machine and the play machine will backspace a certain amount of time to get up to speed before the edit begins to ensure a smooth edit. For example, some systems will backspace 12 seconds, while others will backspace less or more. Both the record machine and the play machine will pause until both are at their designated backspace point. Then both will release the pause and begin the edit. All this happens automatically.

On-screen Display

Time code readings, whether control track pulses or recorded time code, are displayed not only on the machine itself but on the TV/monitor. These readings include the current count as the machine is playing as well as edit data. These edit data are displayed continuously and include the edit-in and edit out points as well as the type of edit, that is, assemble or insert V + LR, V, V + L, V + R, L, R, or LR. This display may also include the source tape, any transition time designated, the duration of the edit, and cumulative edit totals. The basic on-screen display is shown in Figure 7–20. A more sophisticated display is shown in Figure 7–21.

Frame Trim Function

An edit can be trimmed in or back one or more frames at a time. In the basic system, the editor depresses buttons, usually marked + and − to add or delete one frame at a time. In the more sophisticated computer-controlled suite, edits can be trimmed in multiple frames by keying in the number of frames desired.

Swap Control

This switch, shown in Figure 7–22, allows you to use *one* VCR to control a second one. For example, VCR 1 is the play machine in the edit system, and VCR 2 is the record machine. By using the swap control on VCR 1, you can control all the functions of VCR 2 as well, telling it to cue up, pause, play, record, zero the counter, and so on.

Dissolves

A dissolve slowly fades out one picture while fading in another. In effect, the two pictures cross each other. Dissolves require two playback machines and one record machine *and* an edit controller or switcher capable of performing the dissolve and a method of syncing the signals. If you have the ability to do a three-machine edit and have dissolve capability, generally you will also have the ability

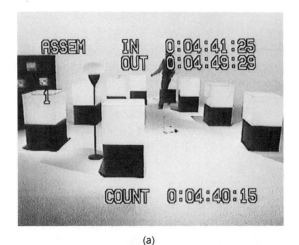

(a)

(b)

Figure 7-20 (a) The on-screen display for an assemble edit. Note the edit will begin at 0:04:41:25 and end at 0:04:49:29, making the edit 8 seconds and 4 frames long. The numbers at the bottom of the screen, 0:04:40:15, indicate where the tape is "parked." (b) The on-screen display for an insert edit. Note that the edit begins at 0:02:35:17 and ends at 0:02:47:27, making it 12 seconds and 10 frames long. Also note that it is an audio edit only, as indicated by the L, meaning left channel audio only. As before, the time code numbers at the bottom of the screen indicate where the videotape is "parked."

to decide how long the dissolve lasts, for example, a 1-second dissolve or a 10-frame dissolve. Generally, a dissolve does not last longer than 5 seconds, and even that may be too long.

Keys

Two pictures are put together to make one in a key. At least one of the pictures must have a chroma key background somewhere in the picture for the key to take place. For example, when the weatherman shows you a map on the weather cast, he is really pointing to a green or blue square on the wall. The weather map is keyed into this green or blue box, as shown in Figure 7–23. What happens is that the camera is told not to see blue or green, so when this color appears, a

Figure 7-21 Some edit controllers have on-screen displays that give specific information on the edit, such as type of edit, transition, and edit-in and edit-out points on the play (source) machine and the record machine, as well as total elapsed time, and so on. (Photo courtesy of FutureVideo Products, Inc.)

hole is created. The second video is inserted into this hole. Anything else in the shot that is the key color may disappear as well. This is why you will sometimes see the weatherman's shirt or suit "keying out," or looking invisible.

Programmable Fades

A fade in or fade out can be programmed into the edit and will be executed automatically when the edit is performed. This is particularly useful at the beginning and end of a program or act, since all traditionally begin with a fade in and end with a fade out.

Figure 7-22 A swap control switch allows you to use *one* VCR to control a second one. This is particularly useful during an editing session.

(a)

(b)

(c)

Figure 7-23 A weathercast, like the one shown here, is an example of a chroma key. The weatherman stands in front of a blue square and points to it. The weather map is keyed into the square electronically. The weatherman sees where he is pointing on the map by looking at the chroma-keyed picture on a TV monitor positioned off camera. (Photos courtesy of Jim Barach and KCOY-TV, Santa Maria, CA.)

Variable-frame Recording

This feature allows you to tell the controller to record a certain number of frames of a scene and then stop recording, and then it will do the edit again everytime you press the *Play* button. The frame record setting varies, but can be as little as 3 frames or as many as 33. This feature allows you to give a scene a time-lapse effect. See Figure 7–24.

Slow or Fast Motion

Some edit controllers allow you to slow or speed the picture by keying in the amount of increase or decrease. Others do not include this as a feature of the machine, *but,* by playing with the jog/shuttle dial, you can usually find at least one setting where you can do this. The pictures from these machines not designed for the feature may lose at least a part of the color and/or may have some scan lines across the picture. However, in order to get the effect, both of these may be acceptable.

Computerized Edit Decision List

Edit-in and edit-out points are stored in the controller's memory with the editor able to delete, add, or rearrange them. If a printer is attached, this list can be printed out and can be used to maintain records of an edit, or it can be used to automatically assemble the program if the system has this capability.

Color Processor Functions

Duplicating some of the features you will find in an SEG, some edit controllers offer color adjustment of the picture. If, for example, the recorded picture is too blue, this feature will allow you to correct the color so that, when it records onto the edit master, it will be the right color. This correction may be displayed in a split screen, so the editor can see the difference between the original recording and the processed one.

Other color processor functions built into some edit controllers provide some special-effects abilities, including the following:

1. *Negative/positive image reversal:* Color pictures can be changed to a negative image, similar to a photo negative.
2. *Monochrome effect:* The color level is reduced, changing the picture into a monochrome picture.
3. *Monotone effect:* A color can be added to the monochrome picture to make it intensely blue or red or green, and so on.
4. *Mosaic effect:* The picture information is digitally processed to produce a

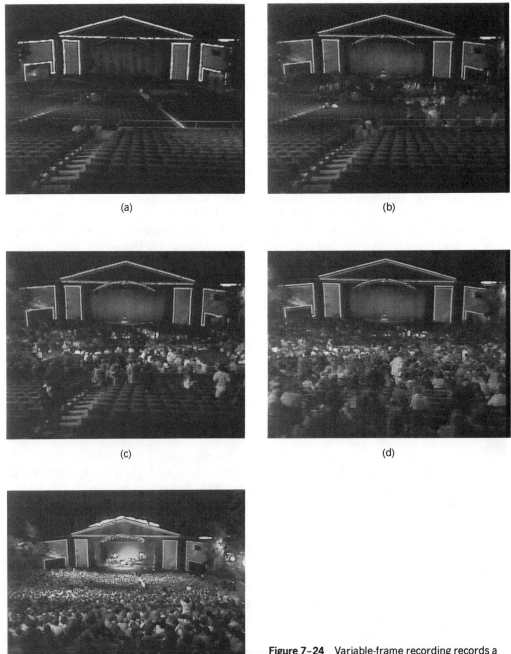

(a)

(b)

(c)

(d)

(e)

Figure 7-24 Variable-frame recording records a certain number of frames, then pauses for a period of time, then records again, then pauses again, and so on, giving a scene a time-lapse effect.

mosaic effect. In other words, the picture is broken down into different sizes of squares, with the effect increasing in intensity until the picture is unrecognizable.

5. *Fade to or from black or white:* When this fade is used, both the video and the audio fade in or out together. The timing of this fade in or out is usually preset, for example, a 3-second fade in.

Character Generator Functions

A limited ability to generate characters for titles and credits may be a part of the edit controller. This ability may include the following:

1. A variety of color backgrounds or the ability to superimpose the characters over a picture
2. Different character sizes (see Figure 7–25)

THERE ARE 24 CHARACTERS
AND 10 LINES AVAILABLE
WITH THIS SIZE TYPE ON
THE JVC HR-S100000U
EDITING SUPER-VHS VCR.
6
7
8
9
10 +‖◄◄►►►&~⊡#✳–|?!)()(',

(a)

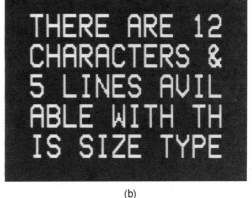

THERE ARE 12
CHARACTERS &
5 LINES AVIL
ABLE WITH TH
IS SIZE TYPE

(b)

THERE AR
E 8 CHAR
ACTERS &

(c)

Figure 7-25 A character generator offers different character sizes of type.

Figure 7-26 A character generator offers different character styles of type.

3. Different character styles, including adding an edge or adding a solid frame (see Figure 7-26)

SYNC AND VIDEO TIMING DEVICES

All multiple video sources must be timed from the same clock in order to switch from one source to another cleanly and without a glitch. This means that every video output must be in sync with all the video outputs of the VCR/VTRs, cameras, SEGs, CGs, and so on. Sync is like marching in a drill team—all must march together. One person out of step throws off the entire team. Some SEGs have a timing circuit built in that will synchronize separate video sources. Others require an external video timing device connected to them to put all the inputs in sync with each other.

There are other video devices designed to improve the picture. Both sync devices and picture-improvement devices process the video in order to make a correction. All video processors should be used judiciously. Everytime the video signal is routed through another electrical circuit, it will lose part of the original clarity, even though the particular device is meant to improve the signal and the look of the picture. The amount of improvement is truly in the eye of the beholder.

Some of the external video processors available are discussed next.

Digital Frame Synchronizer

This device grabs the sync of one video source and the video from a second video source and synchronizes them while in the digital mode. It then converts both to analog in sync and releases them to be recorded on videotape. See Figure 7-27.

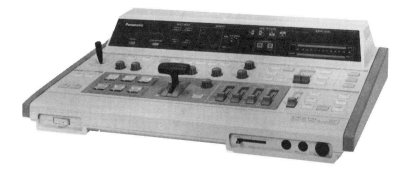

Figure 7-27 A digital frame synchronizer converts analog signals to digital ones, synchronizes all sources, and then converts the signals back to analog before recording. (Photo courtesy of Panasonic Audio/Video Systems Group.)

Genlock

This sync device can be a separate unit or can be built into some character generators, computers, or cameras. This device locks one video signal to another so that the outputs of each are in sync and can be switched in the vertical interval. A separate unit is shown in Figure 7-28.

Time Base Corrector

This electronic device, shown in Figure 7-29, corrects sync problems. Sync problems may cause the recorded video signal to be unstable. There may be tiny

Figure 7-28 A genlock is an electronic device that syncs up various sources. (Photo courtesy of Digital Creations Inc.)

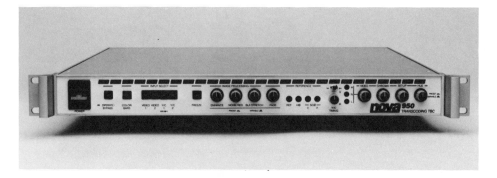

Figure 7-29 A time base corrector (TBC) corrects sync problems that may cause the recorded video signal to be unstable. (Photo courtesy of Nova Systems, Inc.)

movements in the picture detail and the color may waver. As the videotape is dubbed down or edited, these instabilities become more pronounced.

By recording the video through a TBC, you can bring all the different electrical impulses back into sync. The TBC receives the video signal, strips away the previous sync, and replaces it with new sync. This removes some of the time base error. Most TBCs have a limited capacity for the amount of timing correction that they can make.

TBCs are used in the editing process to sync up multiple video inputs. The video should be processed only once through a TBC. For instance, you could process the video when you edit *or* when you dub the edit master, but *never* at both. Some TBCs also have other features, such as the ability to do freeze frames, correct color, and repair dropout.

Processing Amplifier

A video processing amplifier or proc amp, shown in Figure 7–30, separates the different parts of the video signal so that they can be manipulated to make up for possible errors in the original recording. Its main use is to properly set up the video signal on a professionally recorded videotape for dubbing or editing. Some of the things a proc amp does includes (1) corrects vertical and horizontal sync by replacing it with new sync, (2) changes the video level, (3) replaces the color burst if the color is uneven, and (4) processes RGB.

Video Image Detailer

This device makes the picture *appear* sharper by boosting the high frequency. These are the frequencies in the video signal that record the detail part of the

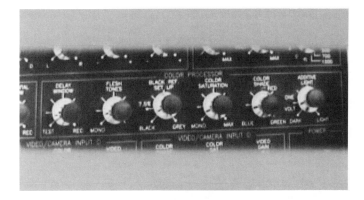

Figure 7-30 A processing amplifier (proc amp) separates the different parts of the video signal so that they can be manipulated to make up for possible errors in the original recording. Proc amps may be separate units or built into other units such as a switcher, as shown here. Note that the proc amp is labeled color processor.

picture. Low frequencies are seen in the smooth, nontextured parts of the picture.

Distribution Amplifier

A video distribution amplifier (DA), shown in Figure 7-31, is used when one video signal needs to be connected to several different input devices, for example, several monitors. It works the same as an audio amplifier, only it amplifies

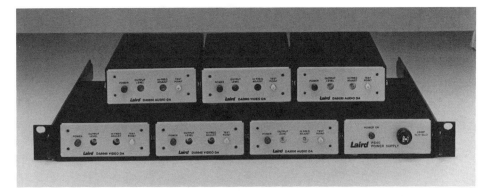

Figure 7-31 A distribution amplifier (DA), such as those shown, amplifies the signal so it can be fed to more than one source, for example, several monitors. (Photo courtesy of Laird Telemedia, Inc.)

Figure 7-32 A character generator (CG) is used to create titles, credits, and other information on the screen. The CG unit pictured is designed for a broadcast edit suite. (Photo courtesy of Laird Telemedia, Inc.)

the video signal instead of the audio. It has one video-in and several video-out terminals. The number of video outs depends on its design.

ADDING A SELF-CONTAINED CHARACTER GENERATOR

Character generators (CGs), shown in Figure 7–32, are used to generate words for titles, credits, and other printed information on the screen. Some also offer other features, such as prerecorded drawings. A CG includes a typewriterlike keyboard for typing in all information. Some of the features you can expect to find on a CG are described next.

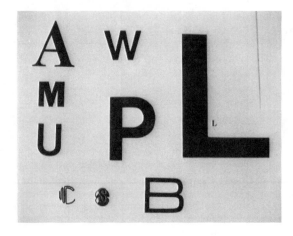

Figure 7-33 CGs offer different type sizes as well as fonts.

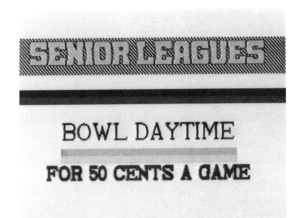

Figure 7-34 CGs offer a variety of type fonts and background colors and designs.

Text On Screen

This is a character generator's main function, to provide wording for titles, credits, and the like, and it will do it in a variety of ways, including the following:

1. Text is available in a variety of sizes and type fonts, as demonstrated in Figures 7–33 and 7–34.
2. The words may be superimposed over a picture.
3. The words may be on a color background, with the picture keyed through the letters.
4. The words may be white, black, or any of a number of other colors.

Figure 7-35 Some CGs will roll the type from bottom to top (shown here) or top to bottom or scroll left to right or right to left.

Figure 7-36 Some CGs will put a drop shadow in a variety of ways around the type they generate.

5. Some will *roll* the words from the top of the screen to the bottom or the other way, and some will *scroll* the words across the screen. See Figure 7-35.
6. A drop shadow on the letters in a variety of forms may be available. See Figure 7-36.

Graphic Designs

A CG may have a number of graphics stored in memory from which the editor can choose, such as one for Happy Birthday or one for Christmas. See Figure 7-37. The most high-tech character generators may have the ability to create basic designs using lines, squares, and circles.

COMPUTER-GENERATED GRAPHICS

Some personal computers have video graphic capability. Some even have the ability to do animation. With a special video card that converts the computer's RGB video to the standard NTSC video signal, a computer can become a video source for an edit session. Computer-generated graphics have the same features that character generators have in terms of displaying on-screen text.

Other features may include the following:

1. Shrinking or enlarging graphics, as shown in Figure 7-38
2. Drawing pictures and, in some cases, animating them, as shown in Figure 7-39
3. Pixelizing the picture, as shown in Figure 7-40

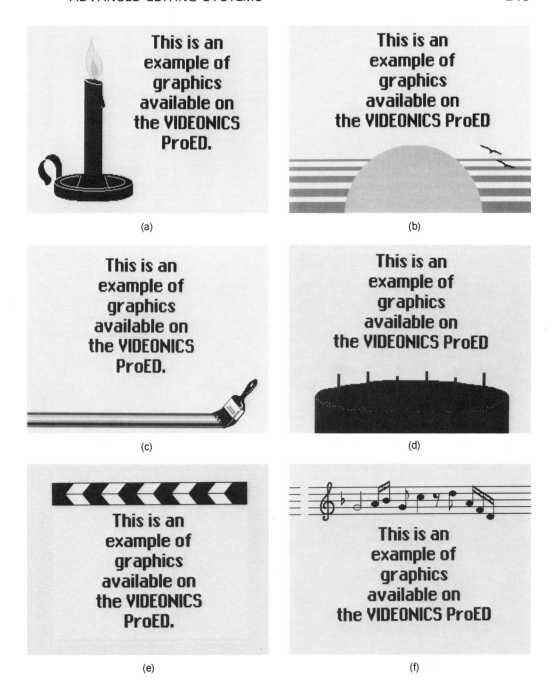

Figure 7-37 A CG may have a number of graphics stored in memory. These graphics can be combined with type generated by the CG. (Graphics courtesy of Videonics.)

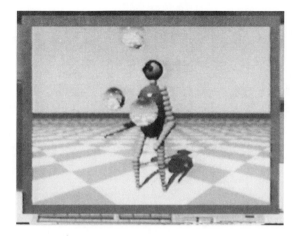

Figure 7-38 A personal computer may have the ability to shrink or enlarge graphics generated by it. In this graphic, the floating balls can be shrunk or enlarged to create an animated effect.

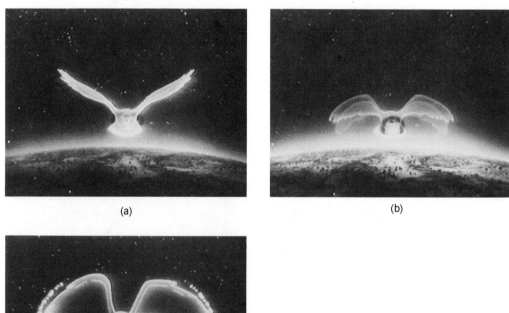

(a)

(b)

(c)

Figure 7-39 Computers cannot only generate graphics but animate them as well, as shown in this series of photographs illustrating a computer-generated and computer-animated eagle.

(a)   (b)

Figure 7-40 A personal computer may be able to pixelize a picture, breaking it down into multiple boxes for effect.

(a)   (b)

(c)

Figure 7-41 Realistic multilayered composite pictures are possible using a colored background, such as blue or green or even red. In the example shown, the girl is shot against a blue background and then keyed over the exterior. In effect, the electronics are told not to see blue in the overlay shot. The result is a shot that makes you wonder if there really are 50-foot women around. (Original color photographs courtesy of Ultimatte Corp.)

4. Generating color bars for reference

5. Creating special effects, such as digital trails, three-dimensional pictures, and keying one picture over another, as shown in Figure 7–41

Computers are bringing home sophisticated editing techniques that until now have been available to only a few with access to high-tech television equipment. *And* this is only the beginning. Where they go from here is anyone's guess.

CONCLUSION

The equipment you can add to an edit system ranges from image enhancers to edit controllers to computers that can create complex special effects *and* control the editing functions. What would you like to have in an edit suite? If you could put your own system together, what features would you want? The obvious answer is "everything you could get." In the next chapter, we will take a look at a basic edit system and the minimum on-line edit suite, utilizing some of the knowledge we have gained up to this point.

TO DO

Figures 7–42 through 7–51 are a series of photos used as illustrations in this chapter. Can you identify each one based on what you have learned?

Figure 7-42 No, this picture is not out of focus. It is actually two scans of the same picture, each positioned slightly off from the other to create what effect? See Figure 7–8 for the answer.

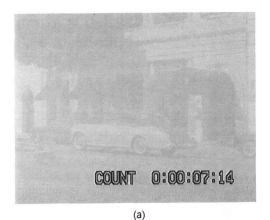

(a)

(b)

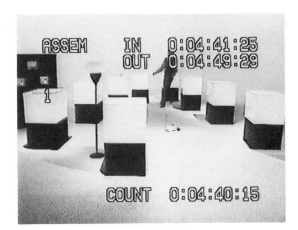

(c)

Figure 7-43 This is an example of a fade up from white. What are the numbers on the pictures? What is the duration of the fade from the first picture to the last according to these numbers? See Figure 7-13 for the answer.

Figure 7-44 The numbers and words at the top of this picture indicate that the machines have been set up to do an assemble edit. What is the duration of the edit according to the numbers? What do the numbers at the bottom of the screen indicate? See Figure 7-20a for the answer.

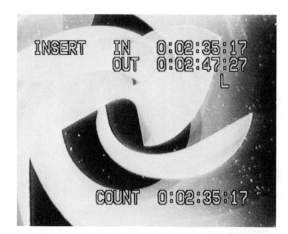

Figure 7-45 The numbers and words at the top of this picture indicate that the machines have been set up to do an insert edit. What kind of insert edit is it? What is its duration? See Figure 7-20b for the answer.

Figure 7-46 This looks like one picture but it is actually two. The weatherman is pointing to a blue background and the map is inserted electronically. This is an example of what kind of effect? See Figure 7-23 for the answer.

Figure 7-47 This an example of an image reversal or changing a picture from a positive to a what? See Figure 7-7 for the answer.

Figures 7-48 through 7-51 Can you identify these four pieces of equipment?

1. Which is a TBC? What does TBC stand for? What does a TBC do? See Figure 7-29 for the answer.
2. Which is a DA? What does DA stand for? What does a DA do? See Figure 7-31 for the answer.
3. Which is a CG? What does CG stand for? What does a CG do? See Figure 7-32 for the answer.
4. Which is an ADO? What does an ADO do? See Figure 7-2 for the answer.

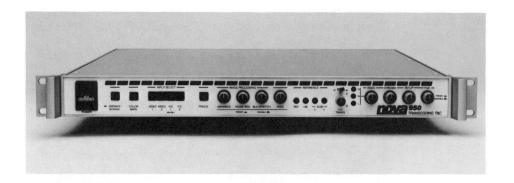

8

Recording and Editing Audio

Audio editing is the most difficult element in the production of a TV program. It is also the last element thought of, if any thought is given to it at all. Yet the sounds are just as important as the pictures, and as much time and effort should be spent on the audio recording as is spent on recording the video. Imagine how your show would be without sound. Turn off the sound and the pictures rarely convey all the moods or the feelings or even tell the complete story. Without the sounds, all you have is part of a TV program.

In comparison to video recording, audio recording is slighted. When we think of TV, we think pictures, not sound. Manufacturers have given more thought to sound as well, but out of necessity. It is simply more difficult to record a video signal than it is an audio signal. The video frequency is 200 times higher than the highest audio frequency and that means that 200 times more information must be recorded to make the picture than is recorded to make the sound. It takes more thought and effort to get the video just right. As a result, the audio becomes secondary.

Manufacturers equip even the most basic camcorder with devices to check the video, like low-light indicators and white balance functions. Most *do not* include any kind of device to tell us when we are in a low-audio situation. We are left to figure out the audio recording on our own by listening to the audio through headphones or watching a VU meter, which displays the strength of the audio signal.

Because it takes this extra effort to check it, audio takes a backseat to video when the original material is recorded. It should not. You would not think of trying to shoot video without looking through the viewfinder and seeing the framing of the pictures, so do not think of recording audio without hearing the sounds as they are being recorded. The quality of the audio is becoming more and more important as sound delivery systems are improved.

AUDIO RECORDING SYSTEMS

Improvements in TV audio did not start to develop until the broadcast industry developed ways to transmit stereo signals over the air that were compatible with all the monaural sets available. Only then did manufacturers start to make TVs with improved sound systems to receive the broadcast stereo signal. In the past several years the major manufacturers of commercial and industrial VCRs, cameras, and camcorders have tried numerous ways to increase the fidelity of the audio signal. The results of their efforts are VCRs that have (1) one linear monaural audio track or two monaural audio tracks; (2) stereo linear audio tracks, called hi-fi on some VCR models; and (3) true hi-fi stereo on the super component video formats that include two normal audio tracks.

Monaural Audio Tracks

Monaural audio can be recorded on one or two audio tracks at the same time. If you are going to record the audio on only one track, use the second track (marked audio 2 or audio R) because it is located farthest away from the edge of the videotape and is thus less prone to edge damage. This is particularly important for reel to reel tape, which does not have the protection of a cassette box surrounding it.

If you have access to two tracks, one can be used to record the dialog portion of your production, while the second track is used to record effects or music. Any sound other than dialog is considered an effect, including sound effects, voice over, or ambient sound. Another option is to use one of the audio tracks to record SMPTE time code.

Stereo Audio

Stereo audio tracks split the audio and record part on the right channel of audio and part on the left channel. These are longitudinal tracks, and the audio is recorded on the length of the videotape. Some VCR/VTRs have two sets of stereo audio tracks, which can almost always be connected separately and recorded on as two monaural audio tracks. Only professional VCR/VTRs have the capability to record or playback true hi-fi stereo audio on linear audio tracks. Consumer and industrial VCR videotape velocity is too slow for hi-fi stereo frequency requirements on longitudinal audio tracks.

Hi-fi Stereo Audio

This type of audio recording is possible with component VCRs that record the hi-fi stereo audio signal *under* the video signal. The disadvantage of this video/audio recording technique is that you cannot edit one without editing the other and maintain the advantage of hi-fi stereo. Machines with this feature provide two linear audio tracks for editing or adding effects or dialog that was not recorded on the original material.

THE PROBLEMS OF PRODUCTION

Some editing problems with the audio elements originate from the time the original recording is shot. The pictures are easy. You can see them in the viewfinder of the video camera or on the field monitor. Monitoring the sounds is more difficult. You use headphones or a VU meter and that is pretty much it. Desperation may drive you to hook up a speaker, but this will only cause problems. The speaker will inevitably be heard by the microphone and become a part of the audio. Also the speaker signal in going through the microphone and the amplifiers creates an endless loop of high-pitched sounds called *feedback*.

Most location problems are caused by the microphones. This is especially true if you are using a camera/camcorder's built-in microphone. If the camera/camcorder is 10 feet away from the main subject, the microphone is 10 feet away also, a bit far to record an adequate audio signal. It is important to plan the audio just as you plan the video. If the camera placement is at a distance, bring in a remote microphone. Do not rely on a camera microphone that is definitely going to be too far away to pick up the audio.

Another problem with location audio is that the sounds that are not important to the story record just as loudly as the important ones; for example, planes fly over, cars go by, and kids play. This can distract the viewer. We are surrounded by sounds; many we are so accustomed to that we do not hear them until they are recorded. Simple things like a camera zoom can add the sound of the lens motor to the tape or a pan away from the subject can cause a lower audio level if the microphone is mounted on the camera. Most audio problems are initially problems of video production, *but* once recorded, they become the editor's problems.

Most audio problems can be resolved in your original plan for the program. Start thinking about audio. Record it right the first time. If an audio signal is weak, the solution is not to turn up the volume or boost the signal in edit; this introduces distortion or muddy sounds. Think of the microphone's sound requirements the same as you think of the video's light requirements and save yourself a lot of problems later in the edit.

Microphone Locations

Where do you position the microphone(s) to record maximum-quality audio? First, remember that, unlike a camera, a microphone can hear only within a certain limited range. It cannot hear as far as the eye can see, as demonstrated in Figure 8–1. Most audio problems have easy solutions, if not obvious ones.

Start by thinking about the obvious—the microphone and its location in relationship to the person speaking. Also consider the kind of microphone in use. Does it hear in one direction only or in all directions? The type of microphone is just as critical as its position.

Types of Microphones

Microphones generate the audio signal by a physical force, that is, the vibrations produced by the sound. The main part of the microphone picks up these vibrations and converts them into electrical energy, which is then transmitted down the audio cable to the recorder. The specific type and design of the mike is designated by the kind of device inside the microphone that converts the vibrations into electrical energy. There are four different types of microphones:

1. *Crystal microphones* use a small crystal to create the signal and are generally the least sensitive and the least expensive.
2. *Dynamic microphones* use a coil of wire and a magnet to generate the signal and are okay for most nonprofessional video recording.

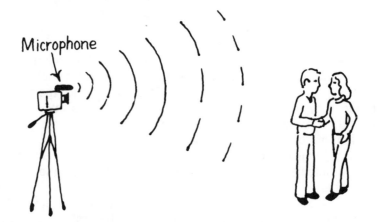

Figure 8-1 Unlike a camera, a microphone can hear only within a certain limited range. It cannot hear as far as the eye can see. (James R. Caruso/Mavis E. Arthur, A BEGINNER'S GUIDE TO PRODUCING TV© 1990, p. 112. Reprinted by permission of Prentice Hall, Englewood Cliffs, New Jersey.)

3. *Condenser microphones* use a condenser to generate the signal and require batteries or an external power supply. This type is more or less the standard for the broadcast industry.

4. *Electret microphones* are an improved version of the condenser mike. Most camcorders have this type, since it can operate with a very small power supply and does not require much space.

Microphone Hearing Patterns

All these types of microphones come in models that hear in different patterns, except the *crystal type*. The five different MIC patterns are as follows:

1. *Omnidirectional microphones* can hear in all directions irrespective of the direction of the sound source. See Figure 8-2.

2. *Unidirectional microphones* are directional, meaning they only hear from one direction, the front. See Figure 8-3.

3. *Shotgun microphones* are also directional type, hearing only from the direction they are facing. See Figure 8-4.

4. *Cardioid microphones* hear very well in the front, not as well to the sides, and hardly at all from the back. See Figure 8-5.

5. *Bidirectional microphones* hear from two directions, front and back. See Figure 8-6.

Figure 8-2 Omnidirectional microphones can hear in all directions irrespective of the direction of the sound source.

Figure 8-3 Unidirectional micro-phones can hear only from one direc-tion, the front.

Just as choosing a microphone with the correct recording pattern is impor-tant, impedance and connectors should also be considered when choosing a mi-crophone.

Impedance and Connectors

Impedance is an electrical measurement in ohms. It is used as a matching desig-nation for *all* audio inputs and outputs, either high or low impedance. In other words, over 1000 ohms is high impedance, sometimes designated as HI Z, and 500 ohms is considered low impedance, or LO Z. *High*-quality microphones are *low* impedance.

Ideally, the *rule* is that "the impedance of the AUDIO OUT should match the impedance of the MIC IN." The impedance matching can be done with an audio-matching transformer if you want to connect a HI Z mike to a LO Z input.

Figure 8-4 Shotgun microphones are directional, hearing only from the front.

Figure 8-5 Cardioid microphones hear well from the front, less well from the side, and hardly at all from the back.

With the impedance-matching transformer, the audio source can send the maximum amount of energy down the line to the recorder. See Figure 8–7.

Here is another *rule* to remember: **"If it works, do it."** And the reason for the rule is that the microphone's impedance must *meet* or be *lower* than the input's impedance. Most record equipment has audio input impedance that is higher than the microphone's output impedance. So if your microphone is low impedance, less than 500 ohms, it will work just fine in most impedance inputs. If it is the other way around, the audio will sound weak and tinny, telling you that you have to match the impedance.

As for connectors, check the connector on the microphone cable to make sure it matches the external mike jack. Most commercial and industrial camcorders, VCR/VTRs, and audio mixers use an unbalanced line (cable) to send

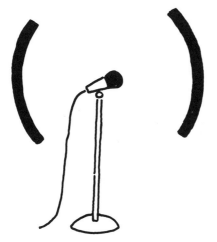

Figure 8-6 Bidirectional microphones hear from two directions, front and back.

Figure 8-7 An audio-matching transformer like this can be used to match the impedance of the audio out to the impedance of the mike in.

the audio signal, including the microphones to the recorder or another device. This line has two wires, the shield (ground) and the center wire that carries the signal. The cable has phone jacks and connectors or the RCA phono type to make the connection to the device. This type of audio or microphone connection may cause problems, picking up stray electrical signals that will degrade your audio signal. In some locations, it will even pick up the transmitter signals from broadcast stations. Sometimes these cables will pick up the 60-cycle hum from the power connections of your editing setup. For this reason it is always a good idea to keep the power cables, your audio, and even the video cables separated from each other.

The other type of audio connector has three wires and three prong connectors and is called *balanced audio*. This type of connector and cable produces a higher-quality audio signal than the unbalanced type and is not as prone to picking up unwanted signals.

If you want to connect a balanced audio source into an unbalanced one, the connector adapter must have a small transformer in it to make the conversion. See Figure 8–7.

Multiple Microphones during Taping

What happens when you have two or more microphones that are set up to record on one audio track on the original material? Or what do you do if you have multiple audio sources, say from a microphone and a CD? It is not as simple as connecting all the audio sources to a splitter and connecting the other end of the splitter to the audio-in terminal jack. Sometimes that will work if both sounds are coming from microphones, but most often it will not work because the two microphones might hear the sounds at different times, causing echo or audio distortion.

The solution is an audio mixer, like the one shown in Figure 8–8. It will allow you to mix multiple audio sources, up to the number of its inputs. The output of the mixer can be monaural or stereo. The amount of audio signal input is controlled for each source so that music and voice can be mixed. The adjustment of the audio level is done with a knob or a fader-type bar called a pot, short for *potentiometer*. This bar regulates the volume of the incoming audio signal before it goes to the output. The number of audio inputs is determined by the capacity of the particular audio mixer. When the output is connected to the audio input of the VCR or camcorder, the combined signal is recorded on the audio track(s).

Ideally, anytime that you have more than one audio source to record on the original material, you should connect them through an audio mixer so that you can control the level of each sound recorded. For instance, if a person is playing the guitar and singing also, you want to be able to hear both the words and the music in a pleasing audio mix without one overpowering the other. With the audio mixer, you can do this because it will allow you to set the volume of each microphone to the most pleasing level for the final recording.

Measuring the Audio

The amount of sound that is recorded can and should be measured when it is being recorded on location or when it is being edited. There is technically an optimum amount of audio signal that should be recorded on the videotape audio track(s) regardless of how loud the sound actually is. For example, if the sound is a bird singing, a person talking, or a jackhammer tearing up the street, there is an optimum amount of sound that should be recorded on most audio systems before the sound becomes distorted. If the sound is recorded below the technical optimum, there is very little that can be done to increase the amount without

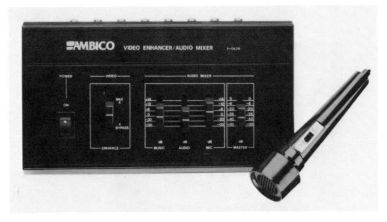

Figure 8-8 Audio mixing boards like the one shown are designed for the consumer editing suite. (Photo courtesy of Ambico, Inc.)

introducing other problems that will distort the sound. If there is too much sound for some audio systems to record, the sound will already be distorted on the audio tracks.

So the next step in recording the optimum amount of the audio signal, *in addition* to using headphones and actually listening to the audio as you are recording, is to measure the amount of sound. You can do this with a VU meter (volume unit meter), which measures the level of the audio signal. Two common types of VU meters are used on video equipment. One is the face and needle type. See Figure 8–9. The other is the LED type. See Figure 8–10.

Both have a **0** (zero) indication on their face. This **0** mark indicates when you are recording the optimum audio level. Below the zero mark are numerical indications that are marked with a minus (−). These numbers are generally black and above the **0**, there are more numbers with plus (+) marks marked in red. The minus numbers like −**20** indicate a very weak audio signal, while a plus number, like +**3**, indicates a very strong and probably distorted audio signal. The needle on the VU meter or the LEDs fluctuate as the amount of sound generated increases or decreases. The optimum recording level is controlled with the pot. As long as the indicator stays in the range of −**3** to +**2**, you are recording the optimum amount of sound.

Another device that might help with some of the problems on the original audio recording is a graphic equalizer. This is an audio processing device that can increase or decrease certain frequencies in the audio signal to make it more pleasing to listen to or enhance the original recording signal.

Kinds of Audio on the Original Footage

Additional audio besides the human voice that you might want to include in the program can be recorded during production as part of the original footage. It is

Figure 8–9   VU meters come in two types. One is the face and needle type. See Figure 8–10 for the other type.

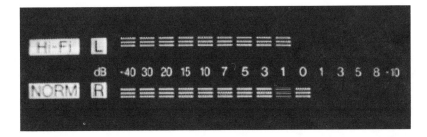

Figure 8-10 The other type of VU meter has an LED readout, in this case, a bar that fluctuates to the intensity of the sound. See Figure 8–9 for a second type.

always a good idea to record natural sounds whenever possible for later use as an editing option or to add to your audio library. Record these sounds the same way you do a voice, placing the microphone in the immediate proximity of the sound source. Always record at the optimum level. You can always lower the volume later when you mix the final track.

Do not forget the noise that is inherently a part of the location—it's *ambient sound.* For instance, rooms have sound, even if it is not obviously audible sound. It is called *room noise,* and it affects the audio that is recorded in the room and how that particular audio sounds to the audience. If this ambient sound is gone, the audio is flat and the scene just does not feel right. The viewer probably will not know why, but will know that something is wrong with the room.

It is these ambient sounds that are most overlooked in recording on location because we are not aware of them. They are part of that particular location even though we do not consciously hear them. For insurance, it is always a good idea to record some ambient sound before you leave a location.

Ambient sound will be important if you are doing a voice-over (VO) video production because it will help the viewer to relate to the locations, particularly if the VO is recorded in a sound booth. Ambient sound mixed under a voice can also be a saver if you want to change the dialog in a production after everything has been recorded. The video in this type of situation can generally be taken care of by cutting to a cover shot while the dialog is changed.

AUDIO MIXING AND EDITING

The easiest audio edit is when every piece of dialog was recorded just where it should be and every sound matches the action in the pictures. The audio and the video cut together sequentially. There is nothing to it. Easy, but not a very practical example since it rarely, if ever, happens. There is always something to add—

a sound effect, music, or a voice over. These combinations of sounds keep the audio exciting and entertaining.

Mixing the Sounds

All the sounds that go into a production regardless of their number must eventually be mixed and recorded on a single audio track for playback. How the sounds are mixed will depend on the number of audio tracks available and how they can be accessed. The different sounds can come from any of a variety of sources, including the original material, an audio cassette, records or CDs, a sound effect library, or anywhere else. A good rule to remember when accessing audio is to *always go direct.* By go direct we mean to connect the audio source directly to the cassette recorder. Do not use a microphone held up to a speaker to record it onto the videotape.

Something else to remember: Audio is more forgiving than video when you need to go down generations. The audio will not degrade as rapidly as the video when dubbed down several generations. For example, if you want four sounds all in the same place to match the pictures, and your audio mixer only has two inputs, you can go down four or five generations on the audio track, generally without a problem.

Here is one way to mix multiple sounds. The first two tracks that are mixed are the ones that are the most important, for example, dialogue and ambient sound. After those tracks are mixed, feed the first-generation audio track through one audio input of the mixer. Then feed the sound effect from the second source to the other audio input. The second-generation audio is recorded with the dialog, ambient, and the sound effects mixed. The original mixed track will be down three generations at this point. Do the entire process one more time, adding the final sound effect, thus taking the original mixed track down another generation to fourth generation.

The successful mixing of the dialog and the effects depends on two factors. The first is the relationship of the volume between the sounds—the mix itself. The levels of all audio sources should be compatible and believable. The second is where those sounds are placed in relationship to the video.

For example, if the visual shows a car crashing, the sounds of a car crash and an observer's reaction must occur at the right moment or the effect will not work. If the sounds of the crash are not timed exactly with the person's reaction, the audience will not believe. On the other hand, if the sound is heard and the reaction is late, the effect will not be believable either. The ideal timing is for the sound to be heard, and then 2 or 3 seconds later, the observer recognizes the event and reacts.

This type of audio control is available in the editing system. Whether it is a basic or a more advanced editing system, you can control the timing of the sounds the same as you control the visuals. The more control the editing system has over the video, the more control it will have over the audio. There is a direct relationship between the two.

Simple One-track Audio Mixing

Let's look at an example of how to mix four sounds onto one audio track. The equipment available for the edit includes the following:

- Two VCRs with VU meters
- One audio cassette player
- One dual audio input mixer with VU meters on each input
- One CD player

One monaural track is available on the VCRs. The audio mixer is a dual-input mixer. There is a VU meter on each input of the audio mixer and one on each of the VCRs. The VO dialog is recorded on an audio cassette. Ambient sound is on videotape; and music and a sound effect will come from compact discs.

Important: Set your record machine to record audio only. On some machines this will be a switch marked audio dub. If you have the ability to do an inserted edit, indicate a L/R edit which means you will record on both the left and right audio channels. If you do not indicate an audio edit, you will record video black, erasing any video recorded on the tape in the record machine. For this exercise, we will be working with a dub of the edit master up until the final edit and we will be putting completely new audio on this dub.

The first step is to record the VO and the ambient sound onto the edit master. The dialog is the most important audio element, and the ambient sounds help give credibility to the location. Connect one input of the audio mixer to the output of the cassette player. Connect the audio out of the play machine to the second mixer input. Connect the output of the mixer to the record machine's audio input. Connect the video out of the play machine to the video input of the record machine.

After all the connections are made and the setup has been tested, put (1) a dub of the edit master into the record machine, (2) the ambient sound videotape into the play machine, and (3) the VO audio cassette into the cassette player. As the audio is mixed, you will be able to hear it through the amplifier of the TV, the monitor, or the headphones connected to the record machine.

Before pressing *Record,* set the levels of all the audio sources. To do this, play the audio cassette player (the VO) and set the pot for its optimum output level by watching the VU meter. Then check the volume of the ambient sound coming off the play machine. The output of the ambient sound will not reach zero on the VU meter so this sound should be mixed by listening to the audio mix either through headphones or over a speaker.

You might want to put a piece of white paper tape along the side of the pot that controls the volume of the record machine, as shown in Figure 8–11, to mark the level for any audio-level changes between the different sources. After the audio sources are set up to your satisfaction on both the VU meter and by listening, set the audio input on the record machine by watching the VCR's VU meter. When it indicates it is receiving the optimum audio signal, you are ready to record your mixed audio track onto a dub of your edit master's audio track.

Now you are ready to begin:

1. Cue up the edit master dub to the edit-in point for the VO and press *Pause.* Then press *Record.* The record machine should now be in the record pause mode.

2. Now cue up the audio cassette (where the VO is recorded) to the edit-in point. Once this is done, press *Play* on the play machine to start the ambient sound. Release the record machine from *Pause;* then release the audio cassette player. You are now mixing the two sounds.

3. Press *Pause* on the record machine when the audio recording is complete. You now have the VO and the ambient sound mixed together onto one track on the edit master dub.

4. The next step is to add the music, which is on a CD, to this mixed track. Pull the cable from the audio cassette player's output and plug it into the output of the CD. Leave the play machine as it was. The setup is shown in Figure 8–12.

5. Put the edit master dub that now has the mixed audio track into the play machine and a work tape into the record machine.

You might want this music to run the full length of the program, but it is more likely that you will want it to cover the open, the close, and some segments

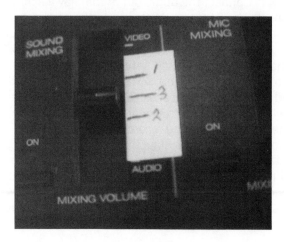

Figure 8–11 Mark audio levels with tape to make sure you record at the right preselected level.

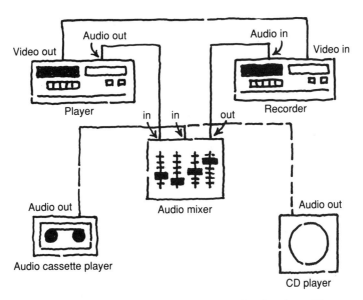

Figure 8-12 This is the setup for a simple one-track audio mix using a CD or a cassette player.

during the program. For this example, let's say you want the music to fade up at first video and then fade under the dialog and then fade out completely. You want to fade up again to cover some scenes and over closing credits. The first thing to determine is exactly where all these fades occur.

If you have an accurate EDL, you will have the counter or time code numbers for inserting the audio. If you do not:

6. Rewind the edit master dub to the beginning and, if you do not have recorded time code, zero the counter. Press *Play* and make a list of where all the audio is to go, noting counter or time code numbers for:
   a. Music audio fades up
   b. Fades down
   c. Fades out completely
7. Next set the different desired audio levels and mark them on the mixer with tape, as before.
8. Cue up the play machine to the first mixed audio track (VO/ambient), the CD to the beginning of the music, and the record machine to the edit-in point.
9. Release the record machine, the play machine, and the CD. You are mixing

the music with the dialog and the ambient sound track. Watch the record machine's counter and fade the music up or down according to your EDL notes. When this step is complete, you have the dialog, the ambient sound, and the music recorded on a single audio track of the worktape. We will call this our audio master.

Now, what about the last sound, which in this case is the sound of a jackhammer? Before you decide how to mix this audio effect, a lot of questions need answers. Some of them might be Why is it there? What do you want it to do? Is it there for effect or is the sound to go with a shot of someone actually using a jackhammer? If the audience is to see a person using the jackhammer, then the questions become How do they see it in the shot? How does the shot start? Is it a pan from something else to the person using the jackhammer? Is it a cut to the person using it?

Then, again, maybe the audience is not supposed to see the person *or* the jackhammer. They just hear the sound. You can see how it might get complex adding a sound like this. How you answer these questions is the basis for the most difficult concept of audio mixing—when and where do we add sound? When we think about the sounds that most things make, we do not usually think of them in the context of how we hear them. Things like how far away the thing making the sound is or even what it is may not even cross our minds.

The concept of hearing the sound itself is difficult because it is abstract. We do not think about how we hear sounds that go with most actions. We are used to hearing words because we watch them coming out of people's mouths. Even talking on a phone or listening to a radio is okay because we know what people look like when they are speaking. We accept a VO narration because it is reinforced by the pictures. We all understand music, and stereo sound or surround sound is accepted even though it is coming at us from different directions because it is reinforced by the visuals. The kind of sound that a jackhammer makes, on the other hand, is on a long list of sounds that requires a thorough understanding of how people hear sounds before they are mixed onto an audio track and exactly what you want to tell the viewer.

The first thing to think about is that the viewer's attention is focused on a small TV screen—small when compared to a movie screen. With the screen size in mind, let's see how we might treat the sound of the jackhammer. We will start by answering some questions. First, the sound is there because it is a show about jackhammers. Second, we want to show how quiet the jackhammer is when it is operating. Third, we want to show how easy it is to use. The video we have edited in already zooms into the jackhammer operator and then pans to the jackhammer.

Having answered these basic questions, we now understand the where and why of the jackhammer sound and it is no longer so abstract. Reduced to its simplest form, we will put it in to match the movement in the pictures. We begin with a WS of the operator.

1. The jackhammer begins at a low audio level because we are on a WS.
2. As the camera zooms in, the jackhammer levels grow louder and louder until, when the zoom ends, it reaches its maximum sound level.
3. As we pan away, the level decreases as we move away from the jackhammer.

What if we changed our minds and wanted to convey a completely different view of the jackhammer. Let's say, "We hate the sound of jackhammers. We would even like to outlaw jackhammers." Visually, we are going to convey this with a cut to a person using a jackhammer and then cut to a person reacting negatively to the sound.

We want to make our viewers sit up and pay attention and to agree with us that the sound of jackhammers is bad. We do this first visually with the abruptness of the cuts as opposed to the slow zoom we did before. We can cut in the jackhammer as well, letting the abruptness of the sound make our point. We could also mix the sound of the jackhammer over the VO dialog so that the sound of the jackhammer obliterates it entirely for a few seconds. The shot of the person's negative reaction to the sound will complete the effect we desire.

To make this final edit:

1. Put the edit master dub into the record machine and cue it up to the WS of the jackhammer operator. Time the sequence, noting when the zoom begins and when it is at its closest. Finally, note when the pan begins and the scene ends. This timing is critical since we will want the sound to be low on the WS and at its loudest when we are closest to the jackhammer. The sound should begin to fade as the pan begins and end abruptly as the scene ends. You will be using these tapes to add this sound hot as the edit is being made.
2. Insert the sound CD into the CD machine. Time the sound, making sure it lasts long enough to cover the scene. If it is longer than you need, select the portion that best suits the scene. Using your notes, practice how you will bring this sound, when it will build to its loudest and when it will cut out. You might want to put tape on the audio mixer to indicate the fader bar position at the entry position, at its highest level and at its lowest level. This will help when you are putting the sound in during the edit. Finally, cue up to the start point selected, set the entry level of the sound and press pause.
3. Put the audio master tape into the play machine and cue it up to first audio at the beginning of the tape. Put the machine in pause at the cue point.
4. Cue up the edit master in the record machine to first video and press pause. The record machine should now be in the record pause mode.
5. Release pause on both the record and the play machine simultaneously. The mixed audio on the audio master will be recorded onto the edit master

dub. Using your notes and watching the video, release the pause on the CD at the beginning of the jackhammer scene. Using the fader bar on the audio mixer, increase the intensity of the sound on the zoom, fade it out as the camera pans away and cut it out when the scene ends.

You now have an edit master dub with a complete audio mix. Dub this audio only over to the edit master and the process is complete. Again, be sure to set your machines to record audio only. You do not want to erase the existing video, only add the mixed audio track.

Two-track Audio Mixing

If your edit system can record on two audio tracks, it will give you a little more flexibility and make the mixing process a little more efficient. In the jackhammer example, you would have recorded the VO on the edit master dub's audio track 1 and the ambient on audio track 2 and then would have mixed the two by playing back the dub from the play machine to record on the work tape in the record machine. One track would then have the mixed dialog, leaving the second track for the effects—the music and the sound of the jackhammer.

If the music and jackhammer were never on at the same time, you would not have to go down another generation at all. You could record the jackhammer sound on audio track 2 and then go back and record the music on the same track. If they are not on at the same time, you will not need to mix them at all.

When you are editing or mixing the audio tracks with the video, be aware of where you set the initial audio level and repeat that identical level at each transition. We want to remind you of this because, no matter how careful you are in the master material recording, differences in the audio recording levels can be recorded onto the audio track. These differences may not be noticeable until the shots are cut together.

CONCLUSION

Audio mixing is just like the rest of the video production—starting at the beginning, through the middle, and to the very end—*it is subjective.* Only you will know when it is right.

Audio is complex. It is much more than what is recorded on the original material. It is music, sound effects, voice over narration and even more, with most of these added during editing. In comparison, the video edit is easy. If you think about audio before the edit, during the taping of the original material, you can save yourself multiple problems with inadequate audio, for example, low levels or unwanted sounds. Plan the audio the way you plan the video, both for the taping and for the edit. After all, picture without sound is only half the story.

TO DO

Here's an exercise to help you test your own system.

1. Look at your VCR. Do you have an audio dub switch? What does the Owner's Manual tell you about it?

2. Hook up two VCRs and one VCR and a camcorder and test them for audio capabilities. Can you add sound to an existing tape using a microphone? Can you use the audio in to plug in another source like a compact disc player?

3. Using a work tape, test the system to see what you can do. Use a work tape with video recorded on it so you can see if the setup is correct for recording audio only. Be sure that the video on the work tape is not important, just in case it gets erased.

9

The Edit Suite

Ideally, the edit suite should include everything you need to do the production the way you visualize it. You should have unlimited access to special effects, graphics, and every transition. You should be able to devote all your energies to making the production the best it can be and not be concerned with whether your equipment can do a particular effect or graphic or transition. That is the way it should be.

Reality is generally a totally different thing. No edit suite is ever perfect. No edit suite ever includes everything, even the most state of the art suites. It is not because they do not have everything, but rather because a good producer and editor can always dream up a new way to do the obvious. The creative is, after all, what producing is all about. Anyone can cut together a program. Not everyone can cut together a great program.

There is then no perfect edit suite, but there are some that come close. In this chapter we describe two that are close: the advanced basic edit suite and the basic broadcast-type edit suite. The advanced basic edit uses control track editing, while the broadcast-type edit suite uses time code editing.

ADVANCED BASIC EDIT SUITE

The most basic of edit systems is counter editing, where the editor must rely strictly on a mechanical counter. There is no frame accuracy and all edits are performed on the fly, with the editor pressing *Record* as the shot plays by. The edits are crude and the production reflects this. A giant step up from this is control track editing.

Control Track Editing

Control track is a series of electronic pulses recorded on the videotape, with one pulse for every frame recorded. All videotape has recorded control track. It is what keeps the tape tracking right as it plays or records. Some VCRs with edit controllers and edit controller units can read control track in time code numbers: hours, minutes, seconds, and frames. The counter is moved forward one frame for every pulse.

Some advantages of control track editing include the following:

Frame accuracy. You can access one frame, if the edit system is capable of doing this.

Real-time counting. The control track reads in real time, so you can time a shot or keep a running total of the length of the edit master.

Some disadvantages include the following:

Slippage by a frame or two. With repeated play, fast forward, and rewind, the tape can slip a frame or two. If this happens, the edit list may be off a frame or two as well.

Beginning zero point may not be consistent. Everytime you take the tape out and reinsert it, it has to be rewound to the beginning and the counter zeroed. If the counter is not zeroed at exactly the same point, accurate to the frame, as it was when you made your edit list, the edit list will be inaccurate. It will probably be close, but it will not be exact to the frame.

The Suite

The basic edit system, as we have learned, is one record machine and one play machine with which we punch and crunch the program together. If these VCRs can read control track as time code, we can improve the system a hundred times, giving us frame accuracy. If these VCRs have a built-in edit controller, all the better. If they do not, then by adding an edit controller, we can improve it even more.

The advanced basic edit suite in its simplest form is available even to the home consumer *and* at a reasonable price. The suite, as shown in Figure 9–1, includes the following:

- One play VCR
- One record VCR
- One monitor
- One edit controller, if it is not part of at least one of the VCRs

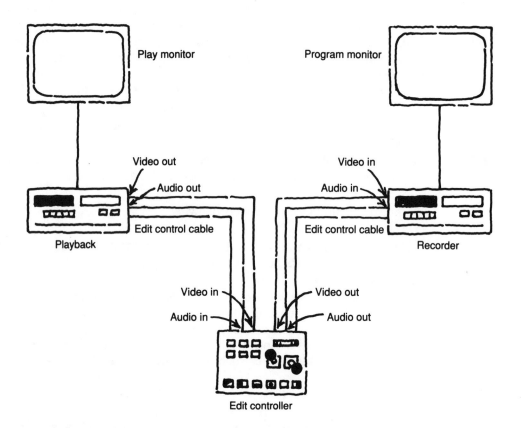

Figure 9-1 An advanced basic edit suite includes two VCRs, at least one monitor, and one edit controller, wired as shown. This setup has two monitors, one for program and one for preview.

With this system, you can do cuts only, control track editing with the ability to do assemble and insert edits with automatic backspacing, fade up/down video and/or audio to black or white, some special effects, and some graphics. This setup also assumes that your edit controller, whether built in or a separate unit, will have a swap control. If it does not, then you should add another monitor so you can see preview on one and program on the other.

From this system, you can add editing capabilities by adding additional equipment. An SEG might be next so that you can add special effects options to your productions. Then perhaps a separate CG so that you can have more variety in font styles and sizes, as well as some graphic options. And, finally, a computer so that you can have the ability to add simple animation to your productions.

A good edit suite is never finished. It is always growing, adding new capabilities and making every production more exciting than the last.

BASIC BROADCAST-TYPE EDIT SUITE

The one major advantage that a basic broadcast-type edit suite has over a basic advanced edit suite is time code editing.

Time Code Editing

Time code editing is 100% accurate because time code numbers are recorded directly onto the videotape. The tape still has control track and it continues to perform its major function—to make sure the tape plays properly. Time code is recorded in one of two places, either (1) in the vertical interval, VITC time code, or (2) on one of the audio tracks, longitudinal time code (LTC). Time code is always SMPTE, the industry standard, meaning it counts the same (hours, minutes, seconds, frames) wherever it is recorded.

Some major advantages of time code editing include the following:

Dependable accuracy. Since time code numbers are recorded on the tape, they will not change, no matter how much you play, rewind, or fast forward. Take the tape out of the machine and put it back in and the time code number *will not* change, ever.

Precise. Because they never change, you can depend on them. If you want to begin an edit in at an exact frame, you can not only find it, you can do it.

Ability to synchronize various sources. Audio can be synced up to video at a precise frame to download an audio mix or add sweetening. Special effects can be added with access to a single frame—the same frame over and over, if necessary.

Time code editing *is* perfect, at least as far as today's standards are concerned.

The Suite

As complex as a broadcast edit suite may be, it is simply an inflated version of the basic suite. Its equipment may be more complex, it is certainly more expensive, and it definitely creates a more technically perfect picture. *But* however elaborate it may be, it begins with the basics. The basic broadcast-type edit suite using time code editing, as shown in Figure 9–2, may include the following:

- One record VTR
- Two play VTRs
- Two monitors, one preview, one program
- Title camera
- Character generator
- Audio board
- Audio amplifier
- Waveform monitor
- Vectorscope

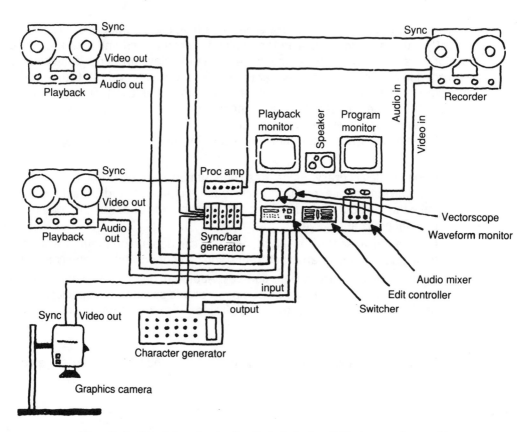

Figure 9–2   A basic broadcast edit suite includes two VCRs, two monitors, a title camera, a character generator, an audio board, an audio amplifier (may be a part of the audio board), a waveform monitor, a vectorscope, a processing amplifier, a TBC, an edit controller and a video switcher. All video processing has a common sync signal.

- Processing amplifier
- Time base corrector
- Editing controller
- Video switcher

Notice that the tape machines have changed from VCRs to VTRs and we have added one play machine. This second play machine gives us the ability to do B rolls, most specifically dissolves. We have also added an audio board and amplifier, which will give us more control of the audio, various sync and picture processing units, and a switcher to put the control of all equipment at one location. In terms of special effects, this basic broadcast system does not have many more, if any, abilities beyond the basic VCR suite. It will, however, have better-quality pictures and audio and it can do dissolves and shoot hard art during the edit. To this basic setup, we might add digital effects or a computer capable of high-resolution graphic animation.

CONCLUSION

There are two basic ways to edit. You can edit using control track, the recorded electronic pulses recorded on the tape, or you can edit using time code, actual hours, minutes, seconds and frames recorded on the tape. Both allow frame access, but control track editing, since it operates on electronic pulses, may change if the tape slips. Time code editing is absolutely accurate because the time code itself is recorded directly onto the tape either in the vertical interval (VITC) or on one of the audio tracks (LTC). The standard for recorded time code is SMPTE time code.

Any edit suite begins with the basics—a record machine and a play machine. To this, another play machine is added, an edit controller, then maybe an SEG, a character generator, a computer, or any number of other machines, all capable of adding something to the edit. A good edit suite is never finished. It continues to grow with new capabilities added or old ones refined. The perfect edit suite is a transitory thing. What is perfect for you today may be insufficient to you tomorrow as your knowledge and capabilities expand and with it your creative thought processes.

TO DO

What kind of edit suite can you create with the equipment you have? Sketch your basic edit suite. Refer to Figure 9–1 as a guide and answer these questions about your suite:

1. How many VCRs will you have? Don't forget you can use a camcorder as a VCR.

2. Do you have an edit controller or are there some edit capabilities built into one of the VCRs?

3. Do you have CG built into a VCR or a camcorder or can you make your computer a part of the suite?

4. Do you have the ability to mix audio without erasing video?

5. Can you connect an outside audio source such as a CD?

Once you have completed your drawing, you should know more about your equipment—what it can and cannot do. Now that you know your capabilities, what is the one thing you would like to add to improve on those capabilities? What one piece of equipment would add the most to the suite? An edit controller? An SEG? An audio mixer? A VCR with built-in control track, frame accurate editing? A CG?

10

How Did They Do That?

This question is asked over and over by viewers. How was a certain effect achieved? Did it require special equipment? Could you do the same thing with the edit suite available to you? Do you really want to? Did the effect add anything to the program or did it distract you? Just because it looks good does not mean it is good. If you remember the effect but forget the program, then the effect did not do its job. It was, after all, a part of the program. It was *not* the program all by itself. In short, an effect should add to the overall effect, not take over.

There is no question that the right effects do add a lot to a program. For this reason, it is always good to know how to perform some, both the basic ones and the fancy ones. Following in Figures 10–1 through 10–20 are just a few effects and how they were done.

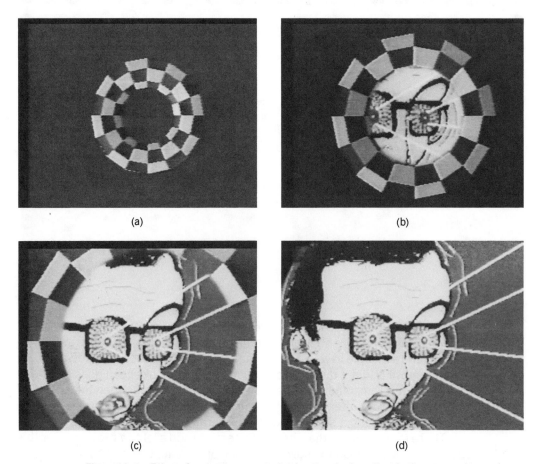

(a)   (b)

(c)   (d)

Figure 10-1   Effect: *Computer-generated animation.* A wipe effect in the same pattern of the glasses worn by the computer character is used as a wipe pattern to go to or from the character, depending on what works best visually with the rest of the production. This effect requires a computer capable of computer animation.

(a)   (b)

(c)   (d)

Figure 10-2 Effect: *Bumper that incorporates the name of the show, as well as interesting, moving visuals.* The final effect is a combination of four video signals that creates the background picture, the two inset pictures, and the graphic. The base or background picture is a straight feed with no effects. The second and third are fed through a two-channel DVE, squeezed down to a smaller size and then inserted over the base picture and moved around the base shot. The graphic, "The Video Game," was fed through the DVE, zoomed, and flipped and recorded on videotape over a black background. It was added to the base picture using a luminance key, which keys off the colors black and white. In the final effect, the base picture plays; the pictures in the two inset boxes play and the boxes move around, even crossing over each other at times; and the graphic flips in the lower-left corner of the screen. This effect requires a two-channel DVE and two playback VTRs.

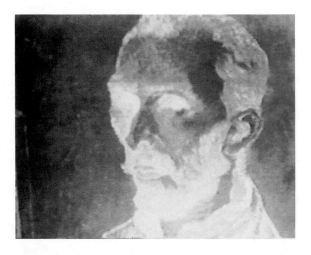

Figure 10-3, 10-4, 10-5 Effect:
*Video signal manipulation to change
the appearance of the picture to add
visual interest.* These three pictures
show three different effects that could
be used in combination or alone. The
first photo is a example of image reversal or changing a positive to a negative.
The second photo (Figure 10–4) shows
random minute pixelization. The third
picture (Figure 10–5) is posterized. All
these effects can be done on a professional switcher and most consumer
SEGs.

Figure 10-6 Effect: *Making a person disappear.* The effect begins with a full screen of the person. This shot is fed to the DVE and squeezed down with a soft-circle wipe around it, getting smaller and smaller until the person is totally gone and only the wipe remains. While the wipe is over the shot, the person leaves the shot. The wipe is then reversed showing that the person has disappeared. This was used on a video game "game show" to fit a video game style. This effect requires a DVE, a VTR, and a switcher with a wipe pattern.

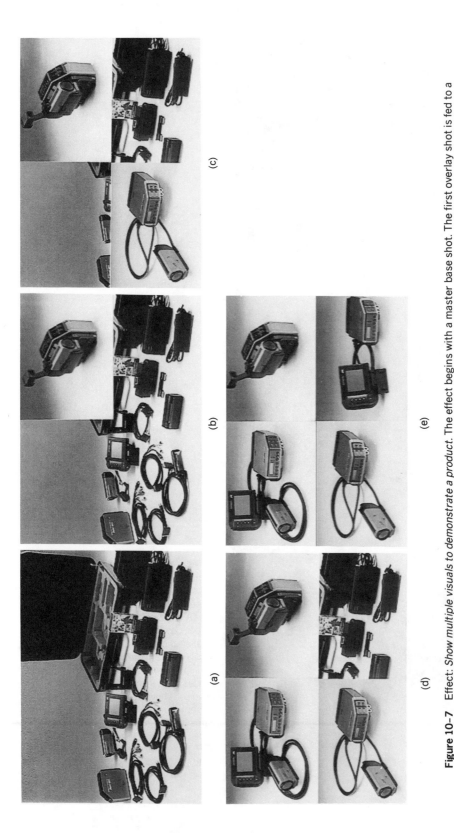

(a)

(b)

(c)

(d)

(e)

Figure 10–7 Effect: *Show multiple visuals to demonstrate a product.* The effect begins with a master base shot. The first overlay shot is fed to a DVE and squeezed down and then slid over the base shot to its position in the upper-right corner. The second overlay shot is fed to a DVE and squeezed and slid into the lower-left corner. The third and fourth shots are added in the same way. The final effect is a quad split with the base picture completely covered. This effect can be done with a four-channel DVE, so it is done in one pass and goes down only one generation. *Or* it can be done with a single-channel DVE and four separate passes. In that case, the first box is recorded on a piece of work tape. That tape is put in the play machine with another work tape in the record machine. The second box is added. Work tape number two is put into the play machine and work tape number one is put in the record machine. The third box is added. Work tape number one is put in the play machine and the edit master is put in the record machine. The fourth box is added. Done this way with a single channel DVE, the first box will be four generations down, the second box three generations down, the third box two and the fourth box one generation. You might start to see picture bleeding at the fourth and third generation.

Figure 10-8, 10-9 Effect: *Bordered shot.* There are two different ways to do a bordered shot. The first shot (Figure 10-8) has been squeezed down to a smaller size using a DVE and the border is added. The second shot (Figure 10-9) has been taped leaving room for a wipe to be put around it. It has not been squeezed. If the shot is squeezed, this effect requires a DVE. If only a wipe is added, this effect can be done with most consumer SEGs.

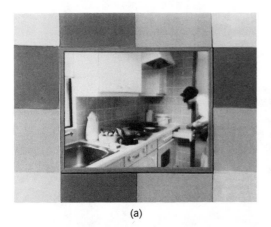

(a)

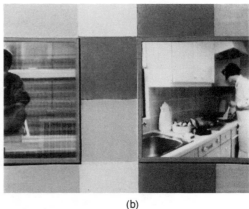

(b)

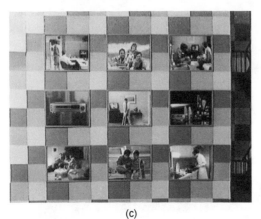

(c)

Figure 10–10 Effect: *Multiple pictures over background ending in a split screen to show all of them.* This effect begins with a computer-generated background. Over this, DVE windows slide across horizontally. Inside the windows is video that has been squeezed to fit by the DVE. This repeats until nine windows have gone by. In the final effect, all nine windows are squeezed and inserted over the background. This effect requires a multichannel DVE. It could be done with a single-channel DVE, but it will have gone done one generation everytime you add a picture, so the last picture would be down ten generations.

Figure 10-11 Effect: *Cut in one picture over another.* The idea is to overlay one picture, but not cut a window for it. In other words, the product appears with no borders around it. To do this, the product is shot in front of a blue or green background. This shot is then keyed through a switcher over the base (city) shot. This effect requires a switcher capable of key effects.

(a)   (b)

Figure 10-12 Effect: *See through the car window.* This effect is created when the picture is recorded using a poli-screen or polarizing filter over the camera lens, turning it to eliminate the reflections in the glass and reveal the driver of the car. This effect requires the camera filter. The effect cannot be done in the edit.

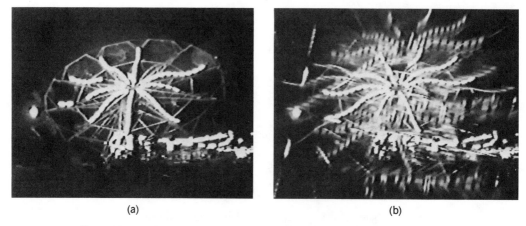

(a) (b)

Figure 10-13 Effect: *Multiple light trailing for emphasis.* This effect is created when the picture is recorded using a multiimage filter. The light trailing is created by the movement of the Ferris wheel. This effect requires the camera filter. This effect cannot be done in the edit.

(a) (b)

Figure 10-14 Effect: *To build four pictures into one.* This multilayered effect is created using four source pictures keyed over one another to make one composite picture. In the first photo, all four pictures are shown stacked. In the second, they are combined into one picture. This effect requires a key source input device, such as the one shown at the bottom of the first picture, or a switcher capable of keying four sources.

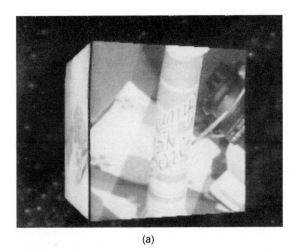

(a)

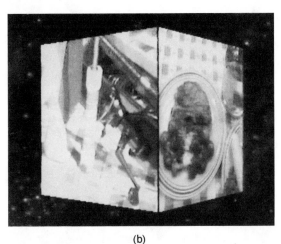

(b)

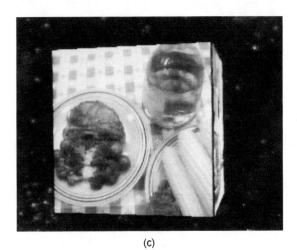

(c)

Figure 10-15 Effect: *Spin a box to show different video on each side.* This effect is an illusion. It is not a spinning box. It simulates a box spinning by squeezing the on-screen video horizontally and vertically and the incoming video horizontally and vertically to create the illusion of a box turning. This effect requires a digital video effects generator capable of squeezing the video horizontally and vertically on the XY axis.

287

Figure 10-16 Effect: *Add sparkle to a shot*. This effect was created using the Ampex ADO. An ADO sparkle effect is overlayed using a half-dissolve over the base picture. This effect can be done with the Ampex ADO or another digital video effects generator with this capability. (Photo courtesy of Ampex Corporation.)

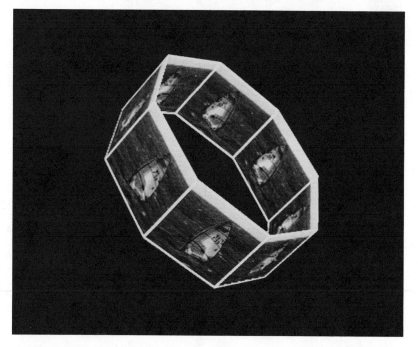

Figure 10-17 Effect: *To create a spinning box*. This is an eight-sided spinning effect created using the Ampex ADO. The ADO takes on video input and multiples it eight times, shaping each to fit the eight-sided pattern and changing these as needed to create the illusion of spinning movement. (Photo courtesy of Ampex Corporation.)

Figure 10-18 Effect: *Add animation to live action.* A computer-generated character is keyed over a live action scene. This effect requires a switcher or SEG capable of keying.

(a)

(b)

(c)

(d)

Figure 10-19 Effect: *Bring in graphics over star background.* This effect starts with a star background. The graphic is fed to a DVE, squeezed down to nothing, keyed over the background, and then expanded until it passes out of the frame. The second graphic follows in the same way.

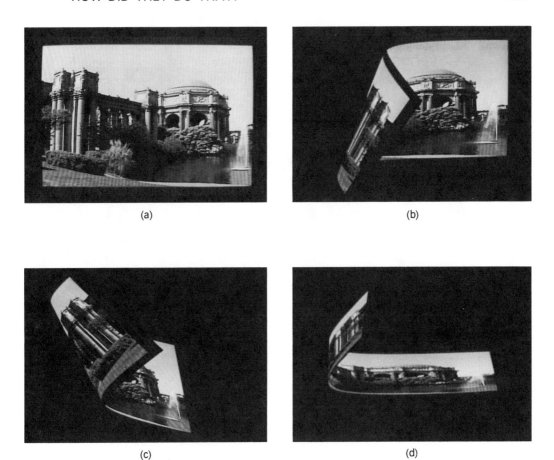

(a)

(b)

(c)

(d)

Figure 10-20 Effect: *3D picture turn.* This effect begins with a black wipe around the picture. This gives the picture more room to turn. Then, with sophisticated digital effects technology, the picture is turned, with the image visible on both the front and reverse of the picture. Also notice that the picture distorts, elongating during the turn effect. This effect was created on an Ampex ADO. (Photos courtesy of Ampex Corporation.)

TO DO

Based on the effects demonstrated in this chapter, how would you do those shown in Figures 10–21 through 10–25? All are another version of the effects shown in Figures 10–1 through 10–20. Read through the chapter again to check your answers.

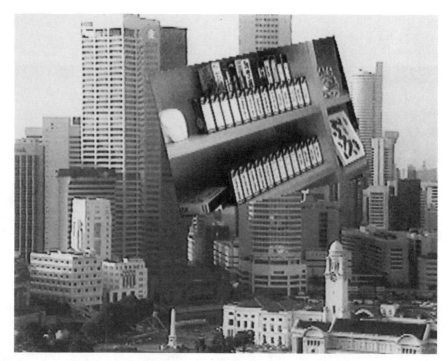

(a)

(b)

Figure 10-21 What kind of effect is this? Assume that the boxes are floating in and settling at a predetermined position on the screen. See Figure 10-7. It is the same type of effect except that the boxes come in differently. What is that difference?

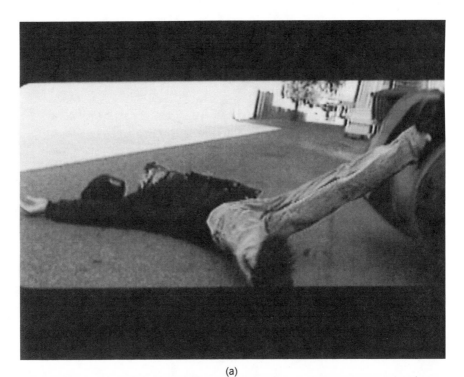

(a)

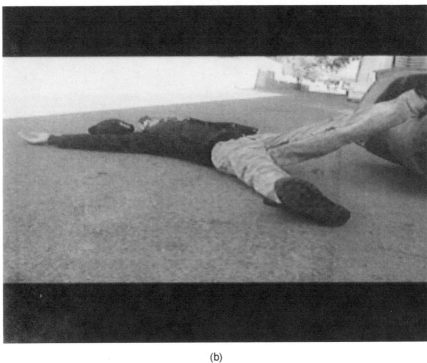

(b)

Figure 10-22 If the first shot is video with a top and bottom screen wipe, what is the second shot? Look at both pictures carefully. There is a difference. Figure 10-8 demonstrates the same effect except that it is a box wipe.

(a)

(b)

Figure 10-23 If the first picture is the original, what kind of effect is on the second picture? See Figure 10-5 for the answer. It has the same effect on it.

(a)

(b)

Figure 10-24 The picture is posterized then duplicated and put into what kind of effect? See Figure 10-10. It has the same kind of effect except that the picture has been split more in this effect, and Figure 10-10 has a background behind the effect.

(a)

(b)

Figure 10-25 The 12 in this effect crawls in from frame right with a wipe following it, revealing the graphic underneath. What kind of effect is the picture portion? Refer to Figure 10-10 for a hint. Which way are the boxes going?

(c)

11

10 Hours to 10 Minutes

At some point in your producing career, you may find yourself with hours and hours of unedited videotape from which you are supposed to create a program. Some of the tapes may have labels. Others may not. Some of them may have some notes about what is on the tape and where to find it and others not. But all the tapes most likely have one thing in common: All were recorded to capture something important, but for one reason or another, the tape was never edited, until now. Now it is your problem.

How do you make a TV program out of this raw video that will tell a story and be entertaining? It is always a real challenge working with existing original material that was shot randomly with no plan in mind for making it into an entertaining short video, but it can be done if you go about it systematically following these 10 easy steps.

STEP 1: SELECT THE SUBJECT

First, start thinking of all that videotaped material as a program. What basic story will it tell? Is it a drama, a comedy, or a documentary? What will it be about? What is the main subject? Make the decision and pick *one* subject.

Keep the subject as broad as you can at this point and do not make it complicated or too limiting. You may discover that you would like to broaden or even change the subject once you have viewed the tape, since there is a definite likelihood that you have forgotten all of what is recorded on those tapes.

Keep it simple. The program will develop as you go through the steps to making 10 hours into 10 minutes.

STEP 2: DECIDE ON LENGTH

Decide how long the production is going to be. Is 3 minutes too long or just right? Maybe a half-hour broadcast time (22:38) is better or maybe the material would make a better fast-paced, 30-second, commercial-length piece. Time is at least one of the major factors to be considered, maybe even the most important. How long can the subject hold the audience? And how much time do you need to cover the subject?

Make a decision based on what you know now. This will give you a goal that you can shorten or lengthen based on what happens between now and the finished program. If you want to make a commercial-length piece, a 30-second commercial is, from fade in to fade out, about 29 seconds, 25 frames or 00:00:29:25 in time code numbers. A half-hour TV program is about 00:22:38:00, which translated means 22 minutes, 38 seconds, zero frames.

If you want to be more flexible, establish a range of time for the length, say 22 to 25 minutes. The object is to determine some finite amount of time that your production is going to run so that you have some place to begin and some end in sight.

If you have a lot of original material or the production is going to cover a long period of real time, consider dividing it into several different episodes like a broadcast miniseries. If you are instructing the viewer on how to do some complicated task, this is also a good approach to lessen the possibility of boring or overtaxing the viewer.

An example to keep in mind when you are making the decision on the length of your program is "Roots"; Alex Haley's classic tale of his search for his African ancestry was told in just 12 hours, including commercials, while the story covered a real-time period of 150 years: 150 years to 12 hours. Ten hours to 10 minutes should be easy by comparison.

The length of your video is important, so important that it is the chief consideration after the basic subject. Why? Because if it is too long, no one will watch, or if it is for broadcast TV, it will not fit into the time slot. In either case, all your efforts will be for naught.

STEP 3: VIEW TAPES AND MAKE SHOT LOGS

The next step is to view the tapes, making a shot log of all material recorded. Use the form provided for copying in the Appendix and seen here in Figure 11–1 or devise a form of your own. It should include the following information:

Show Information

1. *Show title*
2. *Location:* Enter the taping location of the material on the log.

SHOT LOG

Location: _____   Show: _____

Tape No.	Shot No.	Take No.	Time code/counter numbers		Audio	Notes
			In	Out		

Figure 11–1 Use this shot log form to log all material recorded.

Specific Shot Information

3. *Tape number:* Every tape should have a number. Check all the videotapes and give all that do not already have it a number. Write the number both on the tape reel or cassette and on the tape box.

4. *Shot number:* As you view the tapes, give every shot on your log a number. Number them sequentially if they have no numbers at all. Use shot numbers or slates if these are recorded on the tape.

5. *Take number:* The first time you tape a shot, it is take 1. Any subsequent takes are numbered sequentially. You should include on the shot logs all takes under consideration for inclusion in the program.

6. *Time code or counter numbers:* The shot location by time code number or by counter number.

7. *Short audio note:* Is audio recorded? Is it ambient only or does it include important dialog? Is it usable?

8. *Notes:* The framing of the shot as well as a brief description including whether it is an exterior (E) or an interior (I) shot and where it was shot (if there is more than one location on the recorded tape). Framing should be designated as ECU, CU, MS, MCU, WS, OR MWS. This framing information will be important when you plan which shots you will edit together.

Now that you have the shot log form, you are ready to view the videotape. And, yes, you are going to have to look at all of it. What you want to do with this first look is to make a shot log of every tape and get a feel for what is recorded. You might even make some notations about specific shots that are a "must," making the beginnings of an edit list, but this is not your main goal. Your mission is to find out what is on the tape.

If you have not already done so, fill in all the program information at the top of the shot log, which includes title (subject will do if you do not have a definite title yet) and the taping location for this shot log. Before you begin viewing, you will want to make sure that all tapes are rewound. Also be sure to zero the counter on the VCR every time you insert a new tape. If you have time code recorded on the tape, you will not need to bother with this.

Now you are ready to start looking.

1. Put in a tape and rewind to the beginning.
2. Zero the counter, if necessary.
3. Press *Play.*
4. Call the first shot that you see shot number 1 and put this number on the shot log. Also indicate on the shot log the counter numbers or time code numbers where the shot begins and ends.
5. Make a note about the audio.

6. Note the length of the shot, from where you think you would begin using it until the time you think you would cut to another shot. Put this in the Notes column.
7. Now look at the shot, making a note on your shot log about framing, location, and content.
8. Note anything else about the shot that could later be important.

Continue this process through the entire tape, numbering the shots sequentially for easy reference. If some shots are obviously not going to work, you can skip over them entirely. Be careful though. Inevitably, you will decide later you want to use one of these unlogged shots, and it will be time consuming to find it again. Actually, it is a good idea to log everything, just in case.

Look for not only obvious shots but cutaways, establishing shots, shots for effect, montage material, and opening and closing shots. When you finish one tape, start the next. Remember to zero the counter on the VCR when you insert a new tape, if you do not have recorded time code, and to note the tape number on the shot log. If the taped material changes location, start a new shot log. This will make it easier to determine all the shots available at a particular location.

Look at every tape, front to back. In Figure 11–2 is an example of what you might find on the tapes and how it would be logged. In this example, a couple is leaving on a trip. Keep in mind as you are completing your shot log that even though a shot might have been originally recorded as a single 4- or 5-minute shot from start to finish, it can be edited into several shots. You might even want to log it as several shots at this time so that you have a better idea of the framing of every shot.

For example, shots 2–12 through 2–14 (2 is the tape number, 12 is the shot number) together are almost 12 minutes long, covering boarding the plane and getting settled in seats. This could have been logged as one shot. How specific you are in your shot log will be determined by what kind of video you are planning to put together and how important this sequence is to the program. We have logged it as three different shots for the purpose of illustrating the concept of logging shots.

Regarding the length of this particular piece of tape, if the camera operator started recording from the arrival at the airport to the plane taxiing away from the gate without once putting the camcorder in pause, this sequence alone would be almost 2 hours long. If the camera operator only recorded the really important parts using the pause button, it could still run well over 50 minutes. In either case, it is *too* long for TV, besides which the subject is not exciting enough to warrant this much time. Getting to the airport and leaving is only a minuscule part of the story, as we conceive it. To us the real story is where they are going and what will happen to them when they get there. Even so, it could easily be edited down to tell this part of the story in a much shorter amount of time, and we will do that later; but, first, let's look at all the shot logs and decide where we go next.

SHOT LOG

Location: Airport

Show: Trip to Jamaica

Tape No.	Shot No.	Take No.	Time code/counter numbers		Audio	Notes
			In	Out		
1	1		01:01:20:13	01:03:22:13	VO	From inside the car, a WS of the car arriving at the airport.
1	2		01:03:22:13	01:05:00:08	VO	Push to name on the side of the airport main terminal
1	3		01:05:00:08	01:09:00:12	ambient	WS John exiting car, follow to opening trunk/PUSH to CU bags
1	4		01:09:00:12	01:12:09:26	ambient	CU bags/PULL as they come out of the trunk
1	5		01:12:09:26	01:14:11:22	VO	POV bags going onto red cap cart
1	6		01:14:11:22	01:15:14:18	ambient	CU baggage ticket being stapled to ticket by red cap
1	7		01:15:14:18	01:20:18:24	VO	POV walking to security check point John sets off the alarm
1	8		01:20:18:24	01:22:07:17	VO	CU security gun checking John since he set off the alarm
1	9		01:22:07:17	01:27:09:22	VO	POV walk to check-in at gate

Figure 11-2 This is a sample of what a completed shot log might look like.

302

SHOT LOG

Location: Airport

Show: Trip to Jamaica

Tape No.	Shot No.	Take No.	Time code/counter numbers		Audio	Notes
			In	Out		
2	10		02:00:45:22	02:01:11:17	OC	CU airline rep checking us in
2	11		02:01:11:17	02:05:12:02	ambient	WS through window of airplane parked at gate
2	12		02:05:12:02	02:10:15:03	VO	POV boarding the plane
2	13		02:10:15:03	02:11:16:09	ambient	CU airline stewardess at door to plane
2	14		02:11:16:09	02:17:00:21	ambient	POV going to assigned seats
2	15		02:17:00:21	02:18:07:12	OC	CU John
2	16		02:18:07:12	02:19:07:12	OC	CU
2	17		02:19:07:12	02:28:19:23	ambient	WS out the window of taxi away from gate to end of runway
3	18		03:00:19:23	03:05:27:13	ambient	WS out the window of takeoff/MCU
3	19		03:05:27:13	03:08:18:14	VO	Aerial of city through the window of the plane

Figure 11-2 (continued)

When you finish your shot logs, you will have on them the solution to the mystery of what is actually on those stacks of videotape *and,* in finding this out, you will know what your program is going to be about. Also, just from going through it all, you ought to have a good idea of whether you can make the production interesting and whether the time length you originally decided on is too short, too long, or just right.

If you think the estimated length needs changing, wait. You may change your mind again as you pull the usable shots together in the next step.

STEP 4: FINE-TUNING THE CONCEPT

Now that you know what is on your tapes, reanalyze the subject. Are the shots there to make the program you thought you wanted to make? Is there a better idea, given the shots on the tape? Another subject or an entirely different direction that would make a more interesting video?

Look at the title you wrote on your shot logs. Does this still represent the video? Does it still work now that you have seen the tape? If it does, leave it alone. If it does not, change it now. The title at this point is your direction—your guide to pulling all the shots together. Make sure it represents where you want to go with the production.

STEP 5: PULLING THE SHOW TOGETHER

Now that you have settled on the idea and have made your shot logs, it is time to start to develop the idea on paper, to find the sequence of shots that will tell your story. Putting the program on paper for some people is the hardest step in editing. But it is the *most* important step in the total process. Essentially, what you want to do is make a list of shots in the order in which you will put them together. Make this list using your shot logs to locate specific shots. Then review these shots to make sure that they will work, not only in terms of content, both video and audio, but also in terms of editing *from* the previous shot and editing *to* to the shot that follows.

The point is to get something on paper that will tell you if your idea for the program is one that you and the viewer will want to watch after you have edited it all together. Using the shot logs to create this show list, the available shots will stop being just shots and become a part of the total concept of the story, and you should begin to see the completed program in your mind.

Before you begin, you should decide what approach you are going to use. In other words, how are you going to tell the story? Is it going to be told from start to finish in sequence, or will the story best be told in flashbacks? Is it going to be a fast sequence of shots or are you going to be using lengthy shots? Is the audio delivered on camera or is it voice over? Will there be music and will you need sound effects? All these questions will be answered as you begin to create the program on paper.

For now, let's get the basics down by making a list of how we think the program will go together visually, starting at the beginning. Keep the audio in mind as you do this, but concentrate mainly on the video. We will call this show list our scenario and use the form shown in Figure 11–3 (also included in the Appendix for your copying purposes). By preparing the scenario using the shot logs, you will be creating a basic, but rough version of the program on paper. It will give you a place to begin.

STEP 6: FINE-TUNING THE SCENARIO

Now that you have made the scenario, you are just that much closer to creating your program, but you are not quite there yet. First, you have to go back and look at the shots you selected for the scenario one more time. Look for the following:

1. Specific counter or time code numbers for the exact video you want to use
2. Usable audio
3. Exact framing that will work from one shot to the next
4. Places where you will need additional shooting, graphics, CG, voice over, music, or sound effects.

By using the scenario as your guide and treating it like the basis of the story, you will begin to see the completed program come together in your head as you examine tape and choose the final video and audio. Make your notes directly onto the scenario or on a separate sheet of paper as you look, whichever is easier for you.

If a formal script or voice-over commentary is necessary, now is the time to write it, using all the information you have accumulated. Try to keep your script within the bounds of the existing footage so that you do not have a shoot lots of new material. It will not only cost to do this, but it will also slow down the process of creating the program. Your budget and time commitments will probably dictate how far away from the existing material you get in a formal script.

STEP 7: EDIT DECISION LIST

Now, using the scenario, pull the final pieces together and create the edit decision list (EDL), the list we will use when we actually do the editing (sometimes called a paper edit).

The EDL looks a lot like the scenario, but it adds several critical elements: (1) the type of edit: audio only, video only, or audio and video; (2) transitions; and

SCENARIO

For existing master material

Program title _____

Time code/counter numbers

Tape No.	Shot No.	In	Out	Audio	Notes
___	___	___	___	___	_____
___	___	___	___	___	_____
___	___	___	___	___	_____
___	___	___	___	___	_____
___	___	___	___	___	_____
___	___	___	___	___	_____
___	___	___	___	___	_____
___	___	___	___	___	_____
___	___	___	___	___	_____
___	___	___	___	___	_____
___	___	___	___	___	_____
___	___	___	___	___	_____
___	___	___	___	___	_____
___	___	___	___	___	_____
___	___	___	___	___	_____
___	___	___	___	___	_____
___	___	___	___	___	_____
___	___	___	___	___	_____
___	___	___	___	___	_____
___	___	___	___	___	_____

Figure 11-3 The scenario form helps you to put the show together visually in your head by having you make a list of shots as you see them from the beginning to the end.

(3) shot duration. The long form is shown in Figure 11–4 and the short card form in Figure 11–5, both are included in the Appendix for the purpose of copying for continued use.

The edit decision list should have at least the following information:

1. **Scene.** Write here the shot number as noted on the script. This will key this shot back to its place within the script.
2. **Tape number.** Every videotape you will be pulling video and/or audio from should have its own number for easy reference. They should be numbered either consecutively or by program number or date shot or scene number or something that makes sense in terms of the program. *Note:* Time Code is often keyed to a tape number. For example, if the tape is number 6, the time code will designate the hour as 06 (06:00:00:00). This way you immediately know what tape you have up by just looking at the hour on the time code.
3. **Shot number.** This is the number of the shot as designated on the shot log.
4. **Edit type.** There are three kinds of edits: (a) audio only, (b) video only, and (c) audio and video. If your equipment has the ability to access two audio channels, you will need to designate this also, for example, A1 for audio channel 1 and A2 for audio channel 2. If you write only A, then the audio is recording on both channels. What kind of edit are you going to make? Write A or A1 or A2 for audio only, V for video only, and A/V or A1/V or A2/V for audio and video edit simultaneously from the same source.
5. **Transition.** How are you going to get from shot to shot? Will you do it with a cut, dissolve, fade, wipe, special effect, or what? Be sure you know what your equipment can do so that you do not plan a transition you cannot possibly execute. Also, if you are doing a dissolve, fade up/down, or special effect, indicate the transition time. Write in this column one of the following: cut, dissolve/:30 (meaning a 30-frame dissolve), fade up/:10 (meaning a 10-frame fade up), or EFX (meaning an effect). If an EFX, describe the effect in the description column.
6. **Time code numbers or counter numbers.** The exact location of the shot you want to use (in point and out point) or at least as close as you can get, given the equipment with which you are working.
7. **Duration.** How long is the edit? This is helpful for timing purposes. For example, write 5:00, meaning 5 seconds, or 00:10, meaning 10 frames, or whatever.
8. **Description.** Make any notes on the shot, including framing (for example, CU, WS, MCU, narration, effect, 3 shot, shooting comments, and so on.

To complete the EDL, select your shots from the scenario prepared in step 5. Then take one more look at the tape and make final decisions on the exact

EDIT DECISION LIST

Show: _____

Scene	Tape No.	Shot No.	Edit Type	Transition	Time code/counter numbers			Duration	Description
					In	Out	(tape is designated by hour)		

Figure 11-4 Edit decision list long (paper) form.

EDL FOR SHOW

Scene no.	Tape no.	Shot no.	Edit type	Transition
‾‾‾‾	‾‾‾‾	‾‾‾‾	‾‾‾‾	‾‾‾‾

Time code/counter numbers

In	Out	Duration
‾‾‾‾‾‾‾	‾‾‾‾‾‾‾	‾‾‾‾‾

Description

Figure 11–5   Edit decision list single-entry card form.

shots you wanted to use. As you are making the EDL, you will probably discover that you do not have everything you need on tape. It may be as simple as just needing the words for the title, or it could be that you need to tape another shot or that you need a graphic.

Since you did not really plan an edit when you shot the original material, it is certainly likely that you will be missing some element when you go to rearrange it into a program. When you run into this problem, use a line on the EDL to make a note of the missing element, such as "opening title here" or "shot of airline terminal here" or "shot of plane taking off here." To easily locate these additions, make them in a different-colored ink. By doing this, you will know what additional shots, graphics, or words you need to shoot or prepare prior to the edit session.

You may also discover that you need additional audio, voice over, for example, to explain a portion of the video or to further the story line. Whatever you need to make the program complete, make a note on the EDL. It will become your bible—your final source of information.

When your EDL is complete, you are *almost* ready to put the program together.

STEP 8: FILLING IN THE BLANKS

Based on your EDL, you will now need to gather those elements that are not already on tape, like the additional shots you need, the graphics, opening titles, voice over, sound effects, or background music. If there are some postcards, still photos, or slides that will add to your story, now is the time to gather these.

You may or may not want to put some or all of these extra elements on tape. This decision will be based on your editing equipment's capabilities. If you cannot access an audio cassette recorder, record player, or CD, for example, you will want to dub your music and sound effects over to videotape so that you can easily access it. In fact, it is a good idea to put as much as you can on videotape to eliminate additional equipment hookups during the edit and provide easy access to the material.

If you can record voice over directly onto the edit master using a microphone, you might want to do this. If you have a character generator, you will not need to shoot art cards for the title sequence but can access them directly from the machine. In short, now is the time to make all final decisions, because you are approaching the final edit.

STEP 9: ROUGHING IT

At this point you might want to continue to refine the edit on paper to get down to the exact frame in and out points. Or you might want to do a rough cut edit. The idea of the rough edit is to cut the shots together in order without adding other transitions, voice, or graphics. A rough cut can give you a good idea if the concept is working before you spend the money on a final edit.

Whichever method you choose, now is the time to determine how long the program is going to be. You should be able to make a fairly accurate estimate based on your EDL. If you did an actual rough edit, you should be even more accurate. If it is going to run long, now is the time to find out, *before* the final edit.

STEP 10: THE EDIT

Gather all your material: (1) the EDL; (2) all videotapes; (3) all shot logs; (4) any additional graphics, photos, or art cards that have not been recorded; and (5) any dialog, music, or sound that is not recorded yet. You should not have much of item 4 or 5 because you should have recorded most of these onto videotape to make the edit easier.

Once you have all your material, edit your program.

AFTER THE EDIT

Preview your work and make any adjustments. Once you are completely satisfied, it is time to consider the future of the original material. If you are finished forever with these ten tapes, put them in your stash of tapes that can be recycled. If there are some shots on the tapes you want to keep, perhaps as stock footage or for possible future use, dub these shots over to your stock footage library tape. Log the shots and put this tape back on the shelf.

If your setup includes a TBC, you should have no problem with loss of quality going down a generation. If you do not have a TBC, you might want to keep the original tape on which a desired shot is located and save a generation the next time you use it. Make your decision based on the equipment's capabilities and how much you want to keep the shot.

CONCLUSION

It is possible to make a program out of all those tapes in your closet. Perhaps even several programs—and ones that people will want to watch, more than once. You can make the boring interesting by just doing a little editing.

All you need to do is review the tapes, log the shots, plan the transitions, choose any music or sound, decide on any voice overs, and select titles and any other graphics. You may want to shoot some additional shots, to make it all cut together well or you may want to write some dialog to make it more exciting.

Ten hours to 10 minutes is easy; it just takes planning.

TO DO

Using information from the shot log in Figure 11–2, prepare an EDL that will get our travelers in and out of the airport in less than 3 minutes (within the 30–3 rule). Since you are not able to look at the original material, you can assume that all camera angles that you might consider using are there, that is, WS, CU, ECU, MCU, MWS, and so on. Make a copy of the short or long EDL form provided in the Appendix.

Compare your EDL to ours, shown in Figure 11–6. Since our visualization may differ from yours, our EDL may be slightly different. Consider our EDL an example, not the only answer.

EDIT DECISION LIST

Show: _____

Scene	Tape No.	Shot No.	Edit Type	Transition	Time code/counter numbers — In	Out	(tape is designated by hour)	Duration	Description
	1			A	FADE UP				03:52	MUSIC from CD
		1/2	V	FADE UP	01:04:55:08	01:05:05:08		00:10	WS car arriving at airport/PUSH to name on terminal
			A	MUSIC UNDER					
	1	3	A/V	CUT	01:05:30:08	01:05:34:08		00:04	WS John exiting car
	1	5	A/V	CUT	01:13:35:09	01:13:45:09		00:10	POV bags on cart
	1	6	V	CUT	01:14:35:21	01:14:38:21		00:03	CU baggage ticket stapled to ticket
	1	7	A/V	DISSOLVE	01:25:07:14	01:25:37:14		00:30	POV WALK THRU security checkpoint
	1	8	A/V	CUT	01:21:55:17	01:21:57:17		00:02	CU security gun checking John
	1	9	A/V	CUT	01:22:07:17	01:22:27:17		00:20	POV walk toward gate

Figure 11-6 This is a suggested edit decision list for the "Trip to Jamaica" shot log. Use it as an example of an answer for the TO DO exercise.

EDIT DECISION LIST

Show: _____

Scene	Tape No.	Shot No.	Edit Type	Transition	In	Out	Duration	Description
					Time code/counter numbers (tape is designated by hour)			
	2	10	A/V	CUT	02:00:55:13	02:01:00:13	00:05	CU check in at gate
	2	11	V	CUT	02:02:17:22	02:02:37:22	00:20	WS thru window of airplane
	2	12	A/V	CUT	02:07:22:10	02:08:07:10	00:45	POV boarding plane
	2	14	A/V	CUT	02:15:45:12	02:16:00:12	00:15	POV to assigned seats
	2	15	A/V	CUT	02:17:30:22	02:17:35:22	00:05	CU John
	2	16	A/V	CUT	02:17:30:10	02:17:35:10	00:05	CU Mary
			A	MUSIC UP				
	3	18	V	CUT	03:04:35:22	03:05:10:22	00:35	WS takeoff
	3	19	V	CUT	03:06:00:01	03:06:20:01	00:20	Aerial of city

Figure 11-6 (continued)

Appendix

EDIT DECISION LIST: LONG FORM

Show:

Scene	Tape No.	Shot No.	Edit Type	Transition	Time code/counter numbers			Duration	Description
					In	Out	(tape is designated by hour)		

EDIT DECISION LIST: SHORT FORM

EDL for Show

Scene no.	Tape no.	Shot no.	Edit type	Transition
_____	_____	_____	_____	_____

Time code/counter numbers

In	Out	Duration
_____	_____	_____

Description

SCENARIO

For existing master material

Program title _____

Time code/counter numbers

Tape No.	Shot No.	In	Out	Audio	Notes
____	____	____	____	____	_____
____	____	____	____	____	_____
____	____	____	____	____	_____
____	____	____	____	____	_____
____	____	____	____	____	_____
____	____	____	____	____	_____
____	____	____	____	____	_____
____	____	____	____	____	_____
____	____	____	____	____	_____
____	____	____	____	____	_____
____	____	____	____	____	_____
____	____	____	____	____	_____
____	____	____	____	____	_____
____	____	____	____	____	_____
____	____	____	____	____	_____
____	____	____	____	____	_____
____	____	____	____	____	_____
____	____	____	____	____	_____
____	____	____	____	____	_____
____	____	____	____	____	_____
____	____	____	____	____	_____

SHOT LOG

Location: _____

Show: _____

Tape No.	Shot No.	Take No.	Time code/counter numbers		Audio	Notes
			In	Out		

TIMING SHEET

Show _____ Total run time _____

		In	Out
Half-hour show	Opening		
	Commercial 1		
	Act One		
	Commercial 2		
	Act Two		
Hour Show	Commercial 3		
	Closing and credits Act Three		
	Commercial 4		
	Act Four		
	Commercial 5		
	Closing and credits		

Check the appropriate statement:

Time code based on non-drop frame _____

Time code based on drop frame _____

VIDEOTAPE EVALUATION TEST

For: _____

Master Material: _____   Edit master ID: _____

Tape ID	Color Purity	Edge Sharpness	Luminance Capability	Total
_____	_____	_____	_____	_____
_____	_____	_____	_____	_____
_____	_____	_____	_____	_____
_____	_____	_____	_____	_____
_____	_____	_____	_____	_____
_____	_____	_____	_____	_____
_____	_____	_____	_____	_____
_____	_____	_____	_____	_____
_____	_____	_____	_____	_____
_____	_____	_____	_____	_____
_____	_____	_____	_____	_____
_____	_____	_____	_____	_____
_____	_____	_____	_____	_____
_____	_____	_____	_____	_____
_____	_____	_____	_____	_____
_____	_____	_____	_____	_____
_____	_____	_____	_____	_____
_____	_____	_____	_____	_____
_____	_____	_____	_____	_____
_____	_____	_____	_____	_____
_____	_____	_____	_____	_____
_____	_____	_____	_____	_____
_____	_____	_____	_____	_____
_____	_____	_____	_____	_____

Resources

Ambico, Inc.
50 Maple Street
P.O. Box 427
Norwood, NJ 07648-0427
201 767-4100

Amiga Computers
Commodore Business Machines, Inc.
1200 Wilson Drive
West Chester, PA 19380

Canon USA, INC.
One Canon Plaza
Lake Success, NY 11042-1113
516 488-6700

FutureVideo Products, Inc.
28 Argonaut, Suite 150
Laguna Hills, CA 92656
714 770-4416

General Electric Co.
Nela Park
E. Cleveland, Ohio 44112

GTE Products Corp.
100 Endicott Street
Danvers, MA 01923
508 777-1900

Hitachi Sales Corp. of America
401 West Artesia Boulevard
Compton, CA 90220
213 537-8383

322

JVC Company of America
5665 Corporate Avenue
Cypress, CA 90630
714 527-7500

Lowell-Light Manufacturing, Inc.
475 Tenth Avenue
New York NY 10018-1197
212 947-0950

Minolta Corp.
101 Williams Drive
Ramsey, NJ 07446
201 825-4000

Panasonic Industrial Co.
Division of Matsushita Electric
One Panasonic Way
Secaucus, NJ 07094
201 348-7000

Ricoh
180 Passaic Avenue
Fairfield, NJ 07006

Sansui Electronics Corp.
1250 Valley Brook Avenue
Lyndhurst, NJ 07071
201 460-9710

Sima Products Corp.
8707 North Skokie Boulevard
Skokie, Illinois 60077
708 679-7462

Sony Corp. of America
Sony Drive
Park Ridge, NJ 07656
201 930-1000

Steady Ready Media Optics
2606 N. Parish Place
Burbank, CA 91504
818 842-9404

Vertex Video Systems, Inc.
705 Pine Street, Unit E
Paso Robles, CA 93446
805 237-0310

Videonics
1370 Dell Avenue
Campbell, CA 95008-6604
408 866-8300

Zenith Electronics Corp.
1000 Milwaukee Avenue
Glenview, Ill. 60025-2493

Index